T0301819

Leonhard Euler and the Bernoullis

Leonhard Euler and the Bernoullis

Mathematicians from Basel

M.B.W. Tent

CRC Press is an imprint of the
Taylor & Francis Group, an **informa** business
AN A K PETERS BOOK

First published 2009 by A K Peters, Ltd.

Published 2019 by CRC Press
Taylor & Francis Group
6000 Broken Sound Parkway NW, Suite 300
Boca Raton, FL 33487-2742

ISBN-13: 978-1-56881-464-3 (hbk)

Visit the Taylor & Francis Web site at
http://www.taylorandfrancis.com

and the CRC Press Web site at
http://www.crcpress.com

Library of Congress Cataloging-in-Publication Data

Tent, M. B. W. (Margaret B. W.), 1944-
Leonhard Euler and the Bernoullis : mathematicians from Basel /
M.B.W. Tent.
p. cm.
Includes index.
ISBN 978-1-56881-464-3 (alk. paper)
1. Mathematicians--Switzerland--Basel--Biography. 2. Euler, Leonhard, 1707-1783. 3. Bernoulli, Jakob, 1654-1705. 4. Bernoulli family. 5. Mathematics--Switzerland--History--17th century. 6. Mathematics--Switzerland--History--18th century. I. Title.
QA28.T46 2009
510.92'2494--dc22
[B]

2009010076

Cover Illustrations: From left to right, Daniel Bernoulli (p. 196), Jacob Bernoulli (p. 66), and Leonhard Euler (p. 259).

To our friends,
Sabine and Christian Koch

Contents

Preface xi
Acknowledgements xv
Figure Credits xix
1 The Bernoullis as Huguenots 1
2 The Bernoulli Family in Frankfurt and Then Basel 7
3 Jacob Makes His First Steps in the Study of Mathematics 13
4 His Little Brother Johann "Helps" Jacob with Mathematics 21
5 Having Completed His Studies in Philosophy and Theology, Jacob Moves On 27
6 Jacob Travels to Geneva and Meets Elizabeth Waldkirch and Her Family 33
7 Jacob Teaches Elizabeth Waldkirch to Read and Write Numbers and Words 39
8 Sundials, and Tutoring in France 47
9 Jacob Meets with Mathematicians in Paris 53
10 Jacob Travels to Holland and England 61
11 Jacob Settles into Life in Basel to Lecture and Learn 65

12 Leibniz's Calculus vs. Newton's Fluxions 77
13 Johann Bernoulli Grows Up 87
14 Two Curves Studied by the Bernoullis: The Isochrone and the Catenary 97
15 More Mathematical Challenges from the Bernoullis 103
16 Jacob Bernoulli's Mathematics 109
17 Johann Bernoulli Returns to Basel with His Family 117
18 Johann Bernoulli's Son Daniel Grows Up 123
19 Daniel Bernoulli, the Paris Prize, and the Longitude Problem 129
20 Leonhard Euler 133
21 Leonhard Euler's Early Education 139
22 Leonhard Euler Goes to the Latin School in Basel and Then on to the University 143
23 Daniel and Nicolaus Bernoulli Receive a Call to the Academy at St. Petersburg 149
24 The Academy of Sciences at St. Petersburg 157
25 Euler Begins His Career and Moves to St. Petersburg 161
26 Daniel Bernoulli and Leonhard Euler: An Active Scientific Partnership 169
27 The St. Petersburg Paradox 177
28 Euler's Early Work in St. Petersburg 181
29 Daniel Returns to Basel, and Leonhard Euler Becomes Professor of Mathematics at St. Petersburg 191
30 Daniel Bernoulli: A Famous Scholar 201

31 Leonhard Euler: Admired Professor at St. Petersburg 207
32 Euler Becomes Blind in His Right Eye 215
33 St. Petersburg Loses Euler to Frederick the Great of Prussia 219
34 The Eulers Arrive at the Court of Frederick the Great in Berlin 225
35 Euler's Scientific Work in Berlin 237
36 Euler's Work in Number Theory 245
37 Magic Squares 251
38 Catherine the Great Invites Euler to Return to St. Petersburg 255
39 The Basel Clan 263
Index 269

Preface

These mathematicians, who lived between 1650 and 1800, all grew up in Basel, Switzerland. The first two—Jacob and Johann Bernoulli—were important Bernoulli mathematicians who made their careers mainly in Basel. Jacob's name is sometimes given as James in English or Jacques in French, and Johann's name is sometimes given as John in English or Jean in French. Johann's son Daniel, the third great mathematical Bernoulli, spent some years early in his career first in Venice and then in St. Petersburg but returned to Basel as soon as he was able to arrange it. All the other Bernoullis made their careers in Switzerland whenever possible. Only Euler (pronounced "oiler"), who made his career in St. Petersburg and Berlin, chose not to return to Basel. Since the Bernoullis were all related and tended to use the same first names over and over, their names can be confusing, but I hope the family trees in the text will help the reader keep them straight. There is only one Euler who made a career as a mathematician, but he was as important to the development of mathematics as all the Bernoullis taken together. It is unfortunate that most Americans, unless they are crossword puzzle enthusiasts, have never even heard the name Euler. And for that matter, most Americans have heard only of Daniel Bernoulli even though his father Johann and his uncle Jacob were probably equally important.

As I assembled this story, I was disturbed by the minor role played by the women. If the Bernoullis had what might be called the "math gene," surely that was present in the females as well as the males. I assume that the mothers were significant in the upbringing of both boys and girls, although there is also little indication of that influence in the historical record. I think it is likely that Daniel Bernoulli's older sister Anna Catharina was at least partially involved in Daniel and Nicolaus' discussions of mathematics when they were growing up, but that is conjecture on my part. The Bernoulli girls, like other girls at that time, were probably barred from serious education and from later life in the academic world simply because of their gender and the time that they lived. That is regrettable.

The information available on Euler and the Bernoullis is spotty, and in order to make a coherent story I had to fabricate some of the details of their lives and the dialogues that portray their interactions. In general I have tried to convey the interactions of the families and the mathematicians in a way that is compatible with the available records, but there is certainly an element of fiction throughout this work. The quotations from letters are only loose translations, but I have tried to convey both the gist and the mood of the letters. They were written in German, Latin, and French, and I have not made a note of the languages except in one letter that Daniel Bernoulli wrote to Euler, in which he switched repeatedly from one language to another. It is interesting that the correspondents generally preserved the grammar of the disparate languages as they switched from language to language. Their formal letters were all written exclusively in Latin, the language of the scientific community of Europe at the time, while many of their casual letters were written in one or more languages.

Many of the "brilliant but bickering Bernoullis," as William Dunham called them, were indeed cantankerous, particularly when it came to guiding their sons into their careers. For some reason, each succeeding generation apparently tried to force sons into business, law, or medicine rather than mathematics. I have indicated that attitude in part by showing the Bernoulli patriarchs often responding to their sons with a resounding "no!" Leonhard Euler and his father, by contrast, were apparently always kind and supportive as they brought up their children, often responding to the younger generation with a pleasant "yes." It seems to me that that distinction fits with the record, although we have no indication of their use of yes and no.

Another trait the Bernoullis share is that, no matter how cantankerous they were, beginning with Johann they all respected and genuinely liked Euler. That is particularly touching when we consider the contrast between the way the first mathematical Johann Bernoulli treated his son Daniel to the way in which he treated his protégé Euler. Apparently Daniel didn't resent Euler, showing a serenity almost unheard of in a slighted son.

There seems to be general agreement among mathematicians that Euler was one of the four greatest mathematicians of all time, sharing that distinction with Archimedes, Newton, and Gauss. Some have suggested that the whole Bernoulli family should constitute the fifth great mathematician. Among them, these Basel mathematicians had a major impact on the development of mathematics, as well as physics, astronomy, and many other related fields. The two families are certainly responsible for the presentation of Leibniz's calculus to the world, and that alone binds them together.

Since the world may never again see a mathematical clan like Euler and the Bernoullis, it is important that we recognize them for their phenomenal accomplishments and contributions to mathematics. The citizens of Basel didn't ask for a dynasty of mathematicians, but that is what they got. The rest of us can benefit from them as well, but only if we know their story.

Acknowledgments

I want to begin by thanking two remarkable young women who helped me generously in the preparation of this manuscript. Sulamith Gehr, an archivist in Basel, Switzerland, helped me repeatedly, often devoting her precious personal time to tracking down sources for me and later reading my entire manuscript carefully and providing detailed corrections. As we corresponded over the last 18 months, she has never complained about locating the source that I needed and scanning it for me. It is safe to say that without her help this work would be far less accurate and complete than it is. Thank you, Sulamith.

The second young woman whom I want to thank is my daughter, Virginia Tent. While working full time, she managed to find time during her daily subway commute to read the entire manuscript—some parts of it multiple times. Her suggestions showed a real feel for what I was trying to accomplish. On more than one occasion, she urged me to put in more human feeling or to flesh out certain scenes. Her help is particularly memorable on the section where Jacob Bernoulli describes his commitment to mathematics to his reluctant father. The entire book reads better because of Virginia's attentions. Thank you, Virginia.

Next I would like to recognize my two photographers. Lizanne Gray traveled with me to Berlin and Basel in the fall of 2007, taking many, many pictures, both of what I asked her to and what she thought would be appropriate. The result is a wonderful collection of photos that portray many aspects of this story. The lion's share of the photos in this book are Lizanne's work. In addition to Lizanne, my sister-in-law Rosemary K.M. Wyman took two of the photos when I was visiting in Maine. I asked her if she could get a picture of the water flowing under the bridge in the Bagaduce River in Maine and of a snail shell that Virginia Tent found on the shore. Both those photos are masterful. Thank you, Lizanne and Rosemary, for your artistic eyes and technical skill.

I would like to thank my brother, David Wyman, for his help on the work of Daniel Bernoulli. My background in physics is sketchy, but with his knowledge of boats and moving water, David was able to correct my descriptions of navigation and the Bernoulli Principle. It was important that I get those sections right. Thank you, David.

Amanda Galpin, a fine graphic artist, was willing to learn enough about the cycloid to draw its path, depicting a marked wheel as it rolls along a straight path. It is nothing she had ever worried about before, but she approached the challenge directly and quickly, producing what I think is a masterful drawing. Thank you, Amanda.

I needed occasional help in translating some of my sources as well. Although I speak German and French and theoretically read Latin, producing a good English translation of those languages was sometimes beyond my skill level. Jeanne Classé and Jake Linder, teachers of French and Latin respectively at the Altamont School, were repeatedly helpful in fine-tuning my translations. In addition, I should once again thank my daughter Virginia for her help in translating German and French documents. I say to you three, *gratias vobis ago*, *merci beaucoup*, and *danke schön*!

I would like to thank two other archivists in Basel. Dr. Fritz Nagel spent several hours showing Lizanne Gray and me where we needed to go on our walking tour of Basel as we photographed the Bernoullis' environs, and he was most helpful in setting me up for my research in the Bernoulli Archive. Martin Mattmüller at the Euler Archive in Basel was most accommodating as he provided me with sources from his archive as well as a charming paper weight featuring the Leonhard Euler stamp. I particularly appreciate Herr Mattmüller's willingness to send me scans of some documents that I needed to access from Birmingham. Herr Mattmüller's translation of Jacob Bernoulli's poem about infinity is the best that I have found anywhere. Both these archivists provided important material and background information for me. Thank you Dr. Nagel and Herr Mattmüller!

The staff at the Prussian Academy of Sciences in Berlin were most accommodating in providing me with documents and material, and allowing Lizanne Gray to photograph some of their documents. We were particularly charmed with the 1753 almanac, which she photographed in detail. Thank you to the archive staff for their generous help!

Ellen Griffin Shade and Jonathan Newman at the Avondale Branch of the Birmingham Public Library were able several times to help me locate reference materials through their library, often searching for what must have seemed truly bizarre to them. Thank you!

Two of my friends read the manuscript intelligently, giving me some excellent feedback as I revised sections. Mia Cather wanted dates and ages of the characters involved—an excellent suggestion!—and she was also extremely helpful in providing information on her hometown, Groningen, Holland, where Johann Bernoulli served as professor for ten years. Naomi Buklad studied my prose carefully and made several cogent points. Thank you, Mia and Naomi!

At A K Peters, Klaus Peters was supportive and creative in his reactions to my writing. Klaus had a clear vision for this book even when it was in the early stages, and I believe he was right. I sincerely appreciate his comments and suggestions. Charlotte Henderson has always been patient with me, helping me see what I needed to see and providing technical help when I needed it. This book would never have been born without Klaus' and Charlotte's help. Also through A K Peters, Erika Gautschi's copyediting was perceptive and precise. Because she caught several critical errors that I had made in addition to her general editing, this is a far better book than it would have been without her work. I thank you all!

Finally, I would like to thank my husband, James F. Tent. As a professor of German history, he was able to fill in the background that I needed as I wrote—for example, about the persecution of the Huguenots and the role of Peter the Great's Russia in the Europe of the time. Jim also read the manuscript and provided me with important reactions to several sections as I was revising it. I also greatly appreciate that fact that he has supported me in my retirement from teaching, encouraged me, and gone with me in travels to Europe whenever his academic calendar allowed it. Thank you, Jim, as always for your understanding and encouragement.

There are undoubtedly others whom I should mention here, and I apologize to anyone I have omitted. However, I should say that any errors in this book are mine—those who assisted me were wonderful, but I am the one who is responsible for the resulting work.

Figure Credits

Unless otherwise noted below, photographs are by Lizanne Gray and illustrations are by the author.

Page	Credit
11	Rudolph's *Coss*. Courtesy of Bielefeld University Library, http://www.ub.uni-bielefeld.de/diglib/rechenbuecher/coss/.
31	Snail shell. Photograph by Rosemary K.M. Wyman.
98	The cycloid. Illustration by Amanda Galpin.
151	Christian Goldbach. Courtesy of Wikipedia/common Tetra.samlaget.no.
170	Bagaduce River in Maine. Photograph by Rosemary K.M. Wyman.
182	Euler's reciprocal trajectory curve in St. Petersburg Academy journal. Courtesy of Euler-Archiv, Basel.
186	The Bridges of Königsberg problem. Copy courtesy of Euler-Archiv, Basel.
196	Daniel Bernoulli. Courtesy of Bernoulli Archive Basel University.
211	Title page of Johann Bernoulli's *Opera Omnia* [*Collected Works*]. Courtesy of Bernoulli Archive.
231–233	Almanac for the year 1753: exterior, first page, August page, and September page. Courtesy of Berlin-Brandenburg Academy of Sciences.

1
The Bernoullis as Huguenots

"Peter, won't you take some cheese and pass it on?" Francina Bernoulli said to her oldest son as they sat at breakfast one morning in the bustling city of Antwerp in the Spanish Netherlands in the year 1567. "The rest of us are hungry too."

"Oh, Mother!" Peter said, passing the cheese board to his father and taking his first bite. "This is the best cheese!"

"Yes," his mother said, "it's gouda, and it's very fresh. I know you like it best when it is still young, as the cheese maker describes it."

"It's so good!" Peter said enthusiastically.

Francina turned to her husband Jacob, "Did you ever see anyone eat so much?"

"He's a growing boy!" Jacob said. "I remember how hungry I was at his age. By the way, I'll be meeting later today with Justus de Boer. He and I have been exploring working together on shipments of some exotic spices from India. I think it's very exciting."

"I like Justus" Francina said. "I can't think of anyone better to work with."

"No," Jacob said, "I can't either. Not only is he honest and hard-working—he's also smart. You can't ask for more than that in a friend and colleague!"

"Jacob!" Francina Berrnoulli called to her husband as he returned home from work that evening. "Did you hear about Jan Suratt? They burned him alive! Everyone says it is because he refused to acknowledge the Pope! They say the crowd screamed that he was a heretic—that they shouted over and over that he deserved to die!"

"Yes, I heard," Jacob Bernoulli said grimly. "They also burned Justus DeBoer at the stake last night." Jacob sat down at the table and sadly rested his head in his hands.

"Justus? Your friend Justus?" Francina gasped. She quietly put her hand on Jacob's shoulder as together they contemplated the horror of Justus' fate.

"Yes, I know," Jacob said. "Think what my father would have said!" Jacob's father Leon, a devout Protestant, had been a pharmacist and surgeon in Antwerp. He had been one of the leaders in

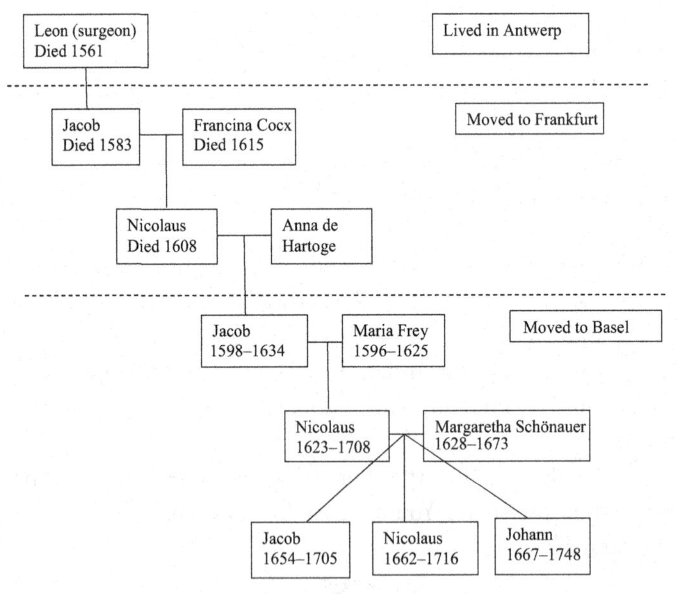

Bernoulli family tree, Antwerp to Basel, 1550–1750.

that exciting port city that was then the thriving center of the Spanish Netherlands' international trade. Leon had been committed to helping his fellow man in every way that he could, and as a surgeon he did what he could to ease the suffering of all people. Among his patients were Protestants and Catholics, Jews and Gentiles, Hollanders and foreigners, and to him the patient's background or religious preferences were irrelevant. He lived the Hippocratic oath: do as much good as possible, but at the very least do no harm. How could the predominantly Protestant city of Antwerp only one generation later have become the scene of deliberate, cruel torture of some of its most respected citizens?

"Jacob, I'm afraid," Francina admitted as she quietly sat down beside him. "The authorities know that we are Protestants, don't they?"

"I'm sure they do," Jacob said. "The Spanish Duke of Alba has made it his business to know such things. He calls us infidels because we have left the Catholic Church."

"Oh, Jacob," Francina said, tears welling up in her eyes, "Do you think we need to leave Antwerp?"

"Yes, I think we should, and I fear we should do it quickly," Jacob said as he shuddered, looking sorrowfully at his wife. Then he continued, "How could they have done this to him? Justus was no threat to them. He wasn't plotting a revolution. He was a good man who always tried to do what was best. He was exactly the kind of man that a civilized mercantile city like Antwerp needs. Why did they care where he chose to go to church—how he chose to worship God? Those are private choices. All people should be able to make those choices for themselves. Oh, dear. Without him and people like him, this center of international trade is nothing. How could they have killed him? It's an abomination!"

"I know," Francina said, taking Jacob by the hand. "He was a fine man."

"Yes, he was," Jacob said. "He was one of the best." Then taking control of his emotions, he continued, "All right, here is what I think we should do: let's go to Frankfurt on the Main River. From what I hear, Protestants are thriving there. It is well known that the Spaniards have no influence in that Free Imperial City of the Holy Roman Empire, so I should be able to continue my business there without fear of persecution. I think it will be best to limit my business there to medicinal spices and drugs, since the diamond trade is best handled from the seaport here. Fortunately for us, establishing the spice trade in Frankfurt is the logical next step in international trade."

"That sounds good, Jacob," Francina said.

"I made inquiries today," Jacob continued, "and I learned that there will be a boat going up the Rhine River from Rotterdam a week from today. I think we should be on it. A carriage would be faster, but because a boat will allow us to take as much as we need, it seems like the best way to go. I spoke today with several of our fellow Protestants, and we agreed that it is best for us to make the move first. You and I will go to Frankfurt with our children first. Because the others are weaker financially, they will have to stretch to make the move. I think it is our responsibility to pave the way for them, and we can do that. If they are cautious and quiet, I hope they won't get caught like Justus and Jan. Once we get established, we can prepare for the others to come as well."

"Yes, Jacob," Francina said. "I think you are right. Your successful business and the money I inherited from my father have set us up well to do this."

"So we will need to leave Antwerp on Monday," Jacob said. "The trip up the river will be very slow—pulling a big boat up the mighty Rhine River is a difficult task—but horses are strong, and they can do it. I hope we will be able to slip away without attracting any notice from the authorities. I'll reserve places for all six of us on the boat."

"I'll start packing at once," Francina said. "Today is Wednesday—we don't have much time! I don't like it, but you are right: we don't have a choice."

"Well, the only alternative would be to convert to Catholicism," said Jacob, "and after what happened yesterday I cannot do that. I am unwilling to submit to the authority of the Pope ever again."

"No, neither of us can do that," Francina agreed. "I will spend tomorrow and the next day sewing gold pieces into the seams of your other shirt and my petticoats. Maybe I can do that to Peter's shirt as well. I think he's old enough for that, don't you? Gold is probably the most portable resource we can take and we have quite a lot, but I will also pack as many clothes for the children as I can. Oh, dear, Jacob! I don't like this at all."

"I don't either, and I agree that sewing gold pieces into Peter's shirt is a good idea. I have some perfect diamonds at the office that you could sew into our clothes as well," Jacob said. "I'll bring them home with me tomorrow. They aren't as heavy as gold, and for their weight they are quite a lot more valuable."

"That's a good idea," Francina said. "Jacob, I'm glad you see it the way I do. I was afraid you might want to stay here and fight. It is appalling that the Duke of Alba is doing this to us!"

"Yes, it is," Jacob agreed, "and perhaps if I were alone I might risk staying here and fighting, but it is unfair to put you and the children in such danger, and yesterday's events prove that the danger is very real. Once we get to Frankfurt, we should be able to prepare the way for all our like-minded friends to come join us, God willing. I pray that they will survive until then."

The Bernoulli family's move was timely. They were able to provide leadership for the Antwerp Protestants in Frankfurt, helping the entire group thrive in their adopted city. Only five years later in 1572, at least 10,000 French Huguenots [Protestants] died in the massacre in France on St. Bartholomew's Day, signaling the beginning of outright war between Catholics and Protestants in Europe. Four years after that in 1576, Antwerp, the primarily Protestant city where the

Bernoullis had lived, was the scene of another cruel slaughter of Huguenots. As many as 8,000 supposed heretics were killed in Antwerp by the troops of the Spanish Duke of Alba on the first day alone, and that included men, women, and even children! After three days, there were no more Huguenots anywhere in Antwerp—they were either dead, or they had escaped with only the clothes they were wearing because of what came to be called the "Spanish Fury." Some had drowned in the river Scheldt after jumping in a final act of desperation. The part of the Netherlands that was under Spanish rule had become a death trap for Protestants, but by now the Bernoullis and their fellow Protestant refugees from Antwerp were thriving in the Free Imperial City of Frankfurt, far from the violence in their native city.

2
The Bernoulli Family in Frankfurt and Then Basel

Frankfurt welcomed the Bernoulli family, and Jacob's business—importing spices from East Asia—was as successful as he had predicted. The family easily made the switch from the Dutch language to German as they adjusted to life in the Rhine-Main region. Jacob and Francina had a total of 17 children although many of them, succumbing to the common diseases of the time, didn't survive beyond their fifth birthdays. By 1570, only three years after his flight from Antwerp, Jacob had become a Frankfurt city councilor because of his impressive success as a businessman. He enjoyed widespread respect in his adopted city. Although at this point a talent in mathematics had not yet been recognized among the Bernoullis, Jacob was clearly able to keep his accounts straight and to make a profit consistently.

Jacob's son Nicolaus continued the family spice business in Frankfurt until 1592, when he moved to the Protestant city of Amsterdam in Holland for a time with his wife Anna. Although he might have wished to return to the family roots in Antwerp, that was not an option. Following the "Spanish Fury," Antwerp had become the most Catholic city in northern Europe, with no tolerance for wayward Protestants. A few years later, Nicolaus returned to Frankfurt to continue the family business there.

In 1620, Nicolaus' son Jacob (grandson of Jacob and Francina who had fled from Antwerp fearing their persecution as Huguenots) decided to move farther up the Rhine River to Basel in what was then called the Helvetian Confederation—what is now Switzerland. With this move, he was removing his family and business from the threats of the emerging Thirty Years War, which ravaged central Europe from 1618 until 1648. By 1622 Jacob, already a well-respected businessman in Basel, was appointed city councilor, probably with some help from the family of his new wife Maria Frey, who was the daughter of a prominent banker and Basel city magistrate. In Basel, the Bernoulli family business in spices continued to prosper.

Jacob and Maria's second son Nicolaus married Margarethe Schönauer, the daughter of a successful pharmacist in Basel, and two of their sons—Jacob and Johann—became the first mathematical Bernoullis, four generations after the family's flight from Antwerp. The mathematical dynasty of the Bernoullis would continue to produce respected mathematicians at an astonishing rate for more than 100 years.

Ever since, mathematicians have argued about whether the Bernoullis had the "math gene"—whatever that might be—or whether each successive generation was somehow brought up to have a passion for mathematics despite their fathers' wishes. Certainly mathematics was never openly encouraged in the family. The "nature or nurture" question in the Bernoulli family is still unresolved, but no one can deny that the family produced at least eight truly great mathematicians within three generations, beginning with the two brothers Jacob and Johann.

In 1668, with the family business now well established in Basel, Nicolaus decided that his very intelligent oldest son—14-year-old Jacob—needn't follow the harried career in business of his father and grandfathers before him.

"Jacob," Nicolaus said to his oldest son one evening, "I have decided that you may be better suited to an intellectual life than to a life in the business world."

"Really?" Jacob asked. "Do you mean that I might study at the university?"

"Yes, I think that would be wise," Nicolaus said. "I've noticed that you are not a fast talker—that you seem to think carefully before you speak. I am almost tempted to say that you seem to have more of a brooding personality—you often seem meditative and deep in thought. What would you think about pursuing a career in the Church?"

"I think I might like that," Jacob agreed. "I must say that a career in business doesn't particularly appeal to me."

"So, I believe what you should do is to study philosophy first," Nicolaus explained, "and then you would move on to the serious study of theology."

"Yes, I like that idea," Jacob said. "In fact, that is what my friend Hans will be doing."

"I'm glad to hear that," Nicolaus said. "You are making me very happy, my boy!"

As directed by his father, Jacob studied philosophy at the university in Basel, and then, after completing his master's degree, he began the study of theology. However, without his father's knowledge, Jacob quietly elected to learn mathematics as well. Since his father expected his children to follow his directions fully, he was furious when he found out.

"Jacob, what is that book you are reading?" his father asked suspiciously one evening.

"It's mathematics, Father," Jacob cheerfully replied. "Most of my reading is in philosophy, but I believe a sprinkling of mathematics is a good balance. Don't you think so?"

"Mathematics?" his father asked. "No! What use could you have for that? Remember, we have reached the point where you can be more than just a businessman. Philosophy is far more important. Since you are a good student, my plan for your career is appropriate."

"But Father," Jacob protested, "You have said that I need to be an educated person, and you must admit that mathematics is certainly part of a broad education. Nothing is as purely abstract as mathematics—not philosophy or even theology."

"No!" his father exploded. "You already know enough mathematics. You learned plenty of that while you were in school, and there is really nothing more to it. You can already do all the reckoning you will ever need to do."

"But Father," Jacob persisted, "I think you don't really understand what mathematics is. It is far more than simple arithmetic. You wanted me to study philosophy, and I have been happy to do that. Plato, one of the greatest philosophers of all time, saw mathematics as the vehicle that draws the soul toward truth. In *The Republic*, his major work in philosophy, Plato argues that the study of mathematics (and by that he means pure mathematics—not just arithmetic) allows one's mind to reach the most ideal truths. He sees mathematics as the perfect vehicle for disciplining the mind. See? My study of philosophy requires me to pursue mathematics, an integral part of that noble subject. I am simply following your directions intelligently."

"Nonsense!"

"That is where you are wrong, Father," Jacob boldly corrected his father. "I have learned that there are some very exciting ideas to be found in pure mathematics, and I have only begun to study them. I would like to understand them all. You wouldn't believe how fascinating it is!"

"That is not what I sent you to the university to study," his father said. "Put that book away!" and with that his father lit a fresh candle, picked it up resolutely, and stormed out of the room.

The book Jacob was studying, which had been published more than 100 years earlier in 1544, was Stiefel's revised version of Christoff Rudolph's *Coss*, an algebra textbook originally published in 1525. The mathematics professor at Basel University had recommended it to Jacob when Jacob asked him what he should read in order to

Die Coſs
Chriſtoffs Rudolffs
Mit ſchönen Exempeln der Coſs
Durch
Michael Stifel
Gebeſſert vnd ſehr gemehret.

Den Innhalt des gantzen Buchs
ſuch nach der Vorred.

Zu Königſperg in Preuſſen
Gedrückt/ durch Alexandrum
Lutomyslenſem im jar
1553.

Rudolph's Coss.

follow Plato's advice and learn more about mathematics. It was the first serious textbook of mathematics beyond basic arithmetic that was available in German, the Bernoullis' language. It presented algebra without the benefit of letters for variables—instead Rudolph used a word (such as the Latin word *facit* [makes] or the German word *gibt* [gives] for our symbol =) or sometimes an abbreviation for a word, to stand for an operation or for the unknown.

Although the mathematics in the *Coss* looks nothing like modern algebra, the *Coss* allowed a student to approach some of the problems found in algebra today, and it was the only way that anyone knew to do algebra at the time. The title *Coss* comes from the Italian word *cosa* [thing], a word that Rudolph sometimes used as his variable. At this time algebraists were often called *cossists*. Jacob had to study the *Coss* seriously if he wanted to pursue his study of mathematics—which he clearly was determined to do.

3
Jacob Makes His First Steps in the Study of Mathematics

Beginning on page 6 of the *Coss*, Jacob found an explanation of series—progressions. He carefully talked himself through the explanation: "All right. First Rudolph presents arithmetic series, in which I should always add the same amount—the common difference—as I move from one term to the next. His first series is the first seven counting numbers—1, 2, 3, 4, 5, 6, 7—in this case I simply add one for each new term. That's easy.

"Now, Rudolph is showing me a trick to find the sum of this series. He says all I have to do is to add the first and last terms—that would be $7 + 1 = 8$—and then multiply the result by the fraction 7/2 to find the total. Now, where did he get that fraction? He must have used seven because there are seven terms, but what about the two? Oh, silly me! Of course! When I add $7 + 1$, I am adding a pair of numbers. In fact there are 3 1/2 or 7/2 pairs of numbers in this series, and each pair must add up to a total of eight. That explains it. I just multiply by the number of pairs. When I multiply 7/2 times eight, that would give me $7/2 \cdot 8 = 28$, and yes, if I add $1 + 2 + 3 + 4 + 5 + 6 + 7$, I get 28. That's good. I like it. Does Rudolph give me another arithmetic series?" Jacob asked himself.

"Yes, the next series is 6, 9, 12, 15," Jacob read. "Now first, I need to be sure that this is an arithmetic series. I see it. There is a common difference of three: $6 + 3 = 9$, $9 + 3 = 12$, $12 + 3 = 15$. That's right. There are four numbers in the series, and, when I add the first and last terms, $6 + 15 = 21$. This time I should multiply the sum of 21 by the fraction $4/2$, since there are four numbers in the series, and so there must be $4/2$ pairs. Since $4/2 = 2$, the total must be $21 \cdot 2 = 42$. That's a good trick!

"Here's another series: 2, 4, 6, 8, 10, 12, 14. The difference between terms is two, there are seven terms, and the sum of $2 + 14$ (the first and last terms) is 16, so I should multiply $16 \cdot 7/2 = 56$. Yes, that's what Rudolph gets, and when I add the terms, that's what I get too.

"Now I believe I understand arithmetic series," Jacob said to himself, "but now Rudolph is moving on to geometric series. I know that with an arithmetic series, there is a common difference between terms, but what about a geometric series? Aha! Instead of adding the same amount from term to term, this time I have to multiply by the same amount. So in the first geometric series on page seven—6, 18, 54, 162, 486—I multiply by three each time, since $6 \cdot 3 = 18$, $18 \cdot 3 = 54$, $54 \cdot 3 = 162$, $162 \cdot 3 = 486$. So the next item in the series would be $3 \cdot 486$ or 1458, a number that Rudolph wants me to find.

"Now, he wants me to subtract six from my new number, 1458. I wonder why. Maybe I should subtract six because the series starts at six. Anyway, $1458 - 6 = 1452$, which he then wants me to divide by two, giving me 726, and that should be the sum of the four numbers. Yes, $6 + 18 + 54 + 162 + 486$ is 726. It gives me the correct answer, but I wonder why. It looks almost like magic, but I'm sure that's not what it is, so there must be an explanation. Rudolph was mighty clever, but I doubt that he was any cleverer than I am.

"Maybe the trick is to divide by the number that is one less than the multiplier—the number that I used to get each of the next terms in the series. This time I multiplied by three, so maybe I divided by $3 - 1 = 2$. That may be the explanation, but I don't have the time now to find out for sure. I think I hear Father coming home for dinner, and I can't let him find me working on this. I hope Rudolph will explain it on the next page. I wish I didn't have to stop now because this mathematics certainly is marvelous! I love it! Rudolph, I'll get back to you and your *Coss* as soon as I can."

A few days later, Jacob was working on a later section of the *Coss*. He found a problem on page ten that Rudolph says Pythagoras might have proposed 500 years before Christ. It was the story of a king who decided to establish 30 cities. For the first city, he would donate one dollar. For the second city, he would donate two dollars. For the third, he would donate four. For the fourth he would donate eight, and so on up to the thirtieth city, proceeding in this way with the powers of two. Today we would say that the first city gets 2^0 dollars, the second city gets 2^1, the third city gets 2^2—with each city getting the number of dollars represented by the power of two that is one less than the number of the city. In this way, the sixth city would get 2^{6-1} or 2^5 or 32 dollars, and so on, all the way up to the thirtieth city, which Rudolph says would require a total that we would write as 2^{29} and that Rudolph wrote as 536,870,912 dollars. However, since at this time the use of exponents was still several years into Jacob's future, he would have had no choice but to multiply by two repeatedly, just as Rudolph had done.

Jacob asked himself, "Is that really the total that I get when I multiply it out? No! It can't be that big! I guess I need to write it out all the way if I want to be sure." Then Jacob continued Rudolph's table. "For the eleventh city, I double the amount for the tenth city: $512 \cdot 2 = 1{,}024$ .... Now, continuing with my doubling, the fifteenth city gets 16,384, or twice as much as the fourteenth city,...." This was getting tedious, but Jacob was determined. "The twenty-ninth city gets 268,435,456, and the thirtieth gets ... Yes, it gets 536,870,912. Remarkable! The amounts started so small, and see how quickly they became enormous!"

Jacob protested, "But these numbers are impossibly big! Pythagoras must have known that no king could have that much money to give to his towns. What a foolish king, and how wise Pythagoras was! Clearly Pythagoras and Rudolph want us to see how incredibly powerful a series of numbers like this can be. How can my father object to this?" Jacob asked himself. "I am supposed to be preparing for the life of an intellectual, and what could be more purely intellectual than mathematics?"

Jacob continued these studies diligently, and within six months he had mastered the *Coss*. What fun it was! And it was so much more exciting to him than pure philosophy! He was developing even greater respect for Plato—the purest of philosophers—who had recognized the purity and importance of mathematics so many centuries ago.

The professor in mathematics at the university knew very little more mathematics than Jacob did by now. His background was in philosophy, but since the position in mathematics was the only one that had been available when he had submitted his application, he had accepted it and had done the best he could. That was common practice at the university in Basel at the time—a professor took a chair in whatever field he could. All university professors had begun with a general philosophical background, many possessing only a veneer of specialization, and many hoped to change into a preferable—or perhaps better paid—field once a better position became available.

The truly great scholars in Europe in the sixteenth or seventeenth century did not make their careers in a university. Instead, they worked in the court of a king or a duke, who expected to derive some prestige for his enlightened court from them and who felt free to ask for an occasional invention or innovation from his resident scholars. By contrast, Jacob's professor at the university was not a great scholar. As was typical at the time, he struggled to handle a heavy teaching assignment, drawing on his limited background but hopeful that perhaps sometime in the future he would be able to pursue a truly intellectual career. As a professor, he was a workingman, condemned to long hours of teaching with only limited compensation.

Fortunately for Jacob, the mathematics professor at Basel was well enough informed to be aware of where Jacob could find some more advanced material in mathematics. He suggested that Jacob look into the writings of Pappus, who had lived in Alexandria on the Egyptian coast of the Mediterranean Sea in the third and fourth centuries A.D. Pappus' work was the most complete presentation of ancient Greek mathematics that was available in Europe at this time. Since the intellectuals of Europe before 1800 revered the Greeks as the greatest scholars ever, Greek mathematics was quite naturally the mathematics they would choose if they were to pursue mathematics.

"But where can I find Pappus' work?" Jacob asked. "Is it in the university library?"

"It should be," the professor replied. "I doubt that anyone has looked at it in many years—the dust is probably very thick on the volume—but the material inside is timeless. The dark ages of early medieval Europe are supposed to be behind us now, but I fear you will be joining a very small group of scholars who will actually be familiar with Pappus."

"Did you find Pappus difficult?" Jacob asked eagerly.

"Oh, I'm afraid I haven't read any of his work," the professor admitted. "I would be surprised if there is anyone in any of the Swiss cantons who has read Pappus."

"But you are the mathematics professor!" Jacob said. "Isn't this supposed to be the oldest and finest Swiss university? How can there be no one on the faculty who has studied mathematics?"

"Most scholars," the professor explained, "are far more interested in philosophy and theology than in mathematics."

"Those are the fields that my father wants me to concentrate on," Jacob admitted, "but I want to do more than that."

"Well, I'm afraid most people share your father's view today," the professor said.

"Do you suppose there might be someone in Geneva who has studied Pappus?" Jacob asked.

"It's possible," the professor said doubtfully. "Since my training was in philosophy, I have read very little mathematics, and I suspect the same is true of the mathematics instructor in Geneva. I can tell you for certain that no one on our faculty is well-grounded in mathematics."

"But why not?" Jacob retorted. "I can't think of anything that is more important."

"I would like to study it," the professor said, "but I simply don't have the time with all the basic courses that I have to teach. Perhaps after you complete your studies you could learn enough mathematics that you could offer the subject more completely than I do."

"I hope that I'll be able to do that," Jacob said.

"It would be wonderful if you could," the professor said.

"Do you suppose that Christoff Rudolph would have read Pappus before he wrote the *Coss*?" Jacob asked.

"I think that is unlikely," the professor replied. "I doubt that it would have been available to him when he was doing his work. I don't think that he could have found the works of Pappus anywhere north of the Alps, and I don't think he ever traveled to Italy. I believe Commandinus' Latin translation of this fourth century Greek work was published in Italy no more than a hundred years ago, and that would have been after Rudolph's time, and it probably wasn't available in any of the Swiss cantons or in Germany even then. I believe that our university library here in Basel bought a copy of Commandinus' Pappus sometime before I became a professor here. At least, it certainly should have bought it."

"Then isn't it strange that I was able to find the *Coss*?" Jacob asked.

"Not really," the professor said. "I suspect that is because the *Coss* is a basic textbook, which has some practical applications in the world of trade. Many businessmen are eager for their sons to be prepared for a life in commerce, and, as you have seen, the *Coss* has some material that businessmen can find useful. Pappus is different. He presents both geometry and logic—it is an interesting combination—with no obvious practical applications. Remember that Plato considered mathematics a part—and he meant an important part—of philosophy. However, I believe you will find it fascinating."

"Thank you," Jacob said, picking up his satchel and preparing to leave the interview.

"*Herr* Bernoulli," the professor added, "I just remembered that there was another mathematician, named Viète—a Frenchman who lived about a hundred years ago—who apparently did some interesting mathematics also. Unfortunately, I know nothing about him, and I have no idea where you could find his work. I have only heard his name. If you can find some of his work, I expect it would be interesting to you as well."

"Thank you. I guess I'll take a look at Pappus' work first if I can find it," Jacob said. "I'll have to wait a bit for Viète since my time is somewhat limited. Unfortunately, I am supposed to be concentrating only on philosophy. However, could you please tell me how to spell Viète's name?"

"He was a Frenchman. I think the French spelling of his name is V-i-è-t-e," the professor said, "but I believe I've also seen it spelled in Latin V-i-e-t-a. He would have written in Latin, of course, and that's the Latin spelling of his name."

"Thank you for the tip, Sir," Jacob said as he bowed politely to his professor and took his leave.

4
His Little Brother Johann "Helps" Jacob with Mathematics

In 1671, Jacob completed his master's degree in philosophy, having put off most of his further studies of mathematics until he had completed that crucial degree. He had satisfied his father by engaging in the expected debates, demonstrating beyond a doubt that he was an informed and articulate scholar of philosophy. The next step was to study theology in order to complete his licentiate in theology, the qualifying course of study for a Reformed minister. However, he had taken the time to find Commandinus' Latin translation of Pappus' work, the *Collection*, in the university library, and now he was ready to tackle it in what spare moments he could find.

Fortunately, his father was not at home this afternoon, so Jacob expected to be able to work in peace. He had the text open on the table in front of him as he was making drawings using a pencil and straight edge (a ruler), carefully following the steps in Pappus' argument. Although the text was accompanied by illustrations, Jacob found that the concepts were easier to follow if he actively constructed them step by step rather than simply looking at Pappus' ready-made drawings.

"Jacob," his four-year-old brother Johann scampered into the room and asked, "what are you doing? Tell me! Tell me! Please!"

"Jacob! Jacob!" Jacob's nine-year-old brother Nicolaus angrily shouted as he stormed into the room at the same moment. "Where did you get that paper? It's mine! Give it back to me."

"Go away!" Jacob said to both his brothers. "I'm trying to work!"

Nicolaus persisted: "Did you take the paper that I left out on the table? Father gave it to me, not to you!"

"I took only a few sheets," Jacob explained. "You still have lots of paper left. Go away and make your pictures. Are you planning to be an artist when you grow up? I can't believe that Father is encouraging you in that."

Nicolaus ran out of the room to see if Jacob had indeed left him enough paper. Jacob had to admit that Nicolaus was pretty good at drawing, although he was surprised in later years when Nicolaus actually became a respected artist.

Jacob then returned to his work, hoping for an uninterrupted hour or two for his studies.

"Jacob," little Johann persisted, "please tell me what you are doing."

"You wouldn't understand," Jacob said. "It's mathematics, and it's a fascinating subject. Since you don't even know how to count yet, I won't bother to try to explain it to you. There is no way you would understand it. Go away, brat!"

"I do too know how to count!" Johann protested. "I can count all the way to 20: 1, 2, 3, 4, 5, 6, 7, 8, 9, 10, 11, 12, 13, 14, 15, 16, 18, 19, 20. See, I'm not such a baby!"

"You skipped 17—it should be between 16 and 18!" Jacob corrected him. "After you have finished learning to count and after you have learned basic arithmetic, I'll teach you some real mathematics, but you'll have to wait a long time for that."

"But Jacob," Johann persisted. "You're drawing something. I can make pictures too."

"No, Johann," Jacob said, "this isn't like Nicolaus' art. It's not just a pretty picture. This is a drawing of Pappus' Theorem. Look

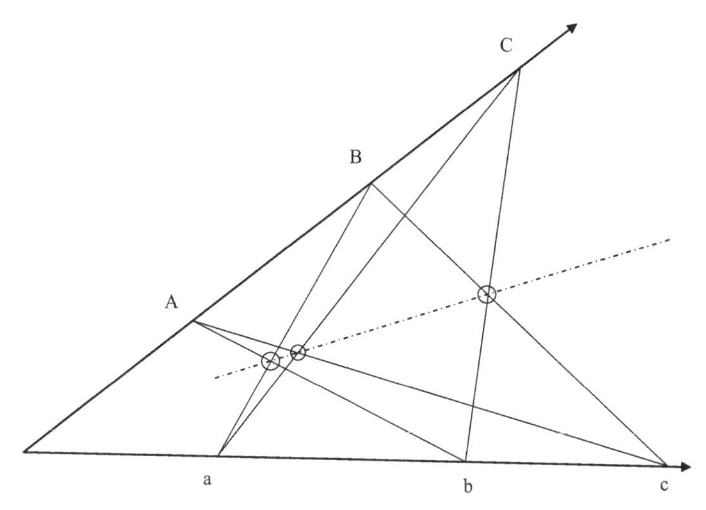

Pappus' Theorem: The three marked intersection points are all on the dotted line.

at what I've done here. See, I have made two straight lines that both start at the same point but go out in different directions from there. On this upper line, I have placed three points that I'm calling capital *A*, *B*, and *C*. On the lower line I have also placed three points, and I'm calling them lower case *a*, *b*, and *c*. You already know the alphabet, don't you?"

"Of course I do!" Johann said. "It's A, B, C, D, …"

"That's enough!" Jacob snapped. "I'm trying to work."

"Is it important to call the points by those letters?" Johann asked. "Couldn't you use other letters—maybe *p*, *q*, and *r*—if you wanted to?"

"I suppose if I wanted to, I could. However, Pappus started at the beginning of the alphabet, so that's where I plan to start too," Jacob explained. "What I'm going to do now is to draw a line from capital *A* to lower case *b* and another line from lower case *a* to capital

B so that I can find the point where those two lines meet. This time I have to be careful to always work with only *a*'s and *b*'s, connecting each capital to the lower case of the other letter. This point that I've marked is the point that I want."

"Are you going to do the same thing with the other letters?" Johann asked.

"That's right," Jacob said. "I'll draw a line from capital *B* to lower case *c* and another line from lower case *b* to capital *C*, this time concentrating only on *b*'s and *c*'s, so that I can mark the point where those two lines meet."

"Why don't you draw a line from capital *A* to lower case *a*?" Johann asked. "You could do that, couldn't you? They're opposite each other too."

"I can't because that's not the way Pappus did it!" Jacob explained impatiently. "Pappus always deals with different forms of two letters at a time. That means that when he's dealing with *a*'s and *b*'s, he takes the capital of one and the lower case of the other, and then he reverses the process: he takes the capital of the other and the lower case of the first in order to locate his point."

"Okay, then what are you going to do next?" Johann asked.

"Now I'll draw a line from capital *A* to lower case *c* and another line from lower case *a* to capital *C* and mark the point where those two lines intersect," Jacob explained. "Careful! You just bumped my arm! Stand back."

"I'm sorry, Jacob," Johann said. "I didn't mean to. I'll try to be more careful. Now what are you going to do?"

"Now I'm going to admire my work," Jacob said. "Look at those three marked points. They are all on a straight line, and Pappus says it will always work out that way. Isn't that amazing?"

"Jacob," Johann asked, "when Nicolaus draws, he doesn't use a straightedge, and he sometimes uses pretty colors."

"I just told you! What I am doing is not art," Jacob explained. "This is mathematics. It is science. I am making a drawing so that I can see what the mathematics looks like. Nicolaus just makes pretty pictures. That is totally different. Mathematics is much more important."

"So what is mathematics?" Johann asked. "Your drawing doesn't have anything to do with counting, does it? I don't see any numbers at all."

"No. What I'm doing is part of geometry," Jacob said, "and geometry is a very important part of mathematics. I've got to work through more of Pappus' argument if I want to understand his proof."

"Do the letters have to be in the same order on both lines?" Johann asked.

"I think so," Jacob said. "I think it matters whether I put capital A, B, and C in one order on their line and then lower case a, b, and c in the same order on their line. Let's try changing the order and see what happens, just to be sure. This time I'll put capital A, B, and C in that order on the upper line, but lower case a, c, and b in that different order on the lower line. Quiet now! I need to do this carefully. Let's see if it works.

"Oh, no!" Jacob exclaimed. "The lines from lower case b to capital C and from capital B to lower case c don't cross when I change the order like that. If they don't cross, I won't have an intersection point to draw the line through. So I guess that shows that the order really does matter—I guess Pappus knew what he was doing. I wonder what happens if I make points capital D and lower case d on my original drawing and work with them the same way I did with capital A, B, C, and lower case a, b, and c."

"Why don't you try it?" Johann cheerfully asked. "Isn't that the best way to find out?"

"Okay, here it is," Jacob said as he continued to draw.

"They look as if they are on a straight line to me," Johann announced.

"Actually, it isn't perfect," Jacob admitted, "but maybe that's because my drawing is not as good as it should be. I think I'll try it again. I guess I have to be careful to always use the very center of each of those points." Jacob concentrated fiercely on his drawing as Johann impatiently waited.

"Yes! They are all on a straight line! You did it!" Johann shouted. "I knew you could do it. My brother, the math man!"

"What I am going to be is a mathematician," Jacob corrected him. "But wait! I'm not so sure that they're all on a straight line. But yes! Yes, I think they are too in a straight line! Look, Johann, if I hold this string over the points and then I pull it tight, all those points are under the string. That means they are all on a straight line."

"I like it, Jacob!" Johann said. "I think it's fun! Will you let me watch you do mathematics again?"

"As long as you don't bother me," Jacob said.

"I was good today, wasn't I, Jacob?" Johann asked.

"Yes, you were pretty good," Jacob said.

"Does Father know what you are doing?" Johann asked in a sweet little voice. "Does he know that you are going to be a mathematician?"

"Of course he doesn't, and you are not going to tell him," Jacob said. "If you breathe a word of this to anyone, I will never allow you to watch me do mathematics again."

"I won't tell," Johann promised. "You can trust me. I plan to grow up to be a mathematician too. Maybe I'll even be a better mathematician than you!"

"Highly unlikely!" Jacob snapped. "I have a head start on you, and all that will be left for you to do is to master what I choose to teach you. Now go away. I have more work to do, and I don't want your help this time. Go somewhere else and practice counting."

5
Having Completed His Studies in Philosophy and Theology, Jacob Moves On

Jacob's father arrived home one evening in 1676 after a long day at work. Only Jacob was around—none of the other children or his wife were at home—and this seemed like a good time for a serious talk with his oldest son.

"What are you planning to do now, Jacob?" his father asked. "You have completed your master's degree in philosophy and your licentiate in theology, you have two calls to become a pastor in the Reformed Church, you are 22 years old, and I think it is time for you to accept one of those calls in the Church. You have already distinguished yourself with two excellent sermons. I am very proud of you. There is no nobler calling than the Protestant ministry, and I believe you are ready for it."

"Oh, no!" Jacob said. "I can't do that—or at least not yet! Well, I suppose I could, but I'm not ready to take that step yet."

"So what do you plan to do?" his father demanded.

"I plan to continue my study of mathematics," Jacob explained. "I already know more mathematics than anyone else in Basel, so I must travel if I want to learn more. I need to find out if anyone in Geneva has studied mathematics. It is an incredibly exciting field, and I must learn more about it!"

"No! What kind of nonsense is that?" Nicolaus asked. "I have already told you that that is not my plan for you."

"But Father," Jacob Bernoulli protested, "I agree that theology and philosophy allow us to approach all of life more thoughtfully and nobly, and I have learned a great deal about them at the university as you wished. But if we stop and think a minute about our family history, you have to admit that working with numbers intelligently and accurately is what allowed our family to become successful importers of spices years ago. Without arithmetic, we would have failed then. What I have learned is that mathematics is far more than adding and multiplying. Just because you don't know anything about it does not mean that it is not important. You will see. I will travel and learn what mathematics has to offer now, and with my knowledge I will take it further than anyone today suspects is possible. I plan to be a great scholar."

"No!" his father said! "That is rubbish, young man! You are arrogant! ... insufferable! It is true that our family has benefited from the arithmetic that has been passed down to us. You are right that it has allowed us to succeed in business, but there is no more to mathematics than that. I am your father, and you will do as I say."

"No, Father," Jacob said. "You must admit that our family has always survived by our wits—our wits strengthened by our knowledge and our integrity. Of course we need to have a firm moral foundation as well as knowledge of our culture, but if we are no more than moral people, we will lose out in the end. Remember, when your great-great-grandfather Jacob left Antwerp, he took a big chance. His father might not have approved of it, but clearly it was the right thing to do. You have to admit that his move to Frankfurt could have been disastrous. Our family's later move to Basel was chancy as well. Those earlier Bernoullis took enormous risks. Father, with all due respect, I would like to take a chance as well, and I believe the result will be similarly good."

"No, Jacob," his father Nicolaus responded, shaking his head sadly. "Certainly our ancestors' move from Antwerp and later from Frankfurt were wise moves, and I don't deny that arithmetic helped our family to establish a solid business. I suppose I have to admit that we have become one of the prominent trading families in Basel, at least in part because of our mastery of arithmetic. I never said that calculating is not important—of course it has helped us—but I am determined that you will have the life that I was not able to have.

"Your grandfather and his grandfather before him fought for our religious freedom. You will be the first in our family to pursue the life of the cloth, and you have completed the studying that you need in order to do that. It makes me proud to think of that. And a life in the Church will be well enough paid that you will be able to support yourself and a family comfortably. That is arithmetic that I can understand very well."

"But Father," Jacob said, "that is not what I want to do—at least not yet. Please allow me to travel to Geneva and then to France so that I can pursue mathematics. Just because you don't understand it does not mean that it is not important, and remember that we are talking about my life—not yours. Many of the men whom I have been studying with are going to travel for a couple of years before they settle down for their life work. While I am traveling, you may be sure that I will take advantage of opportunities to preach so that I will continue to build up a good reputation as a cleric as well. I promise you that I will make you proud before I am done."

"Well, I guess you may take a little more time before you settle down," his father said, "so long as it doesn't interfere with your real career in the Church."

"So you have decided to allow me to learn more about mathematics?" Jacob asked.

"You are trying to trick me into taking your side," Nicolaus barked.

"I must study mathematics. I must travel," Jacob Bernoulli informed his father. "I have the university degrees that you required me to get, but I am not willing to stop there."

"Poppycock!" Nicolaus Bernoulli fumed as he sat down at the table, pounding his fist as he continued to speak. "I can't see that your mathematics will have any application to your life in the ministry. And if you think that you would be able to support yourself and a family with a career in the university, you are wrong. Professors are the poorest of the poor. A parish priest has a far more comfortable life, earning more than twice as much as even the most famous university professor. You've seen them. They have a miserable existence. I have better plans for you."

"In fact, a pastor earns only half again as much, not twice as much," Jacob corrected his father, "but regardless, I must learn more mathematics. The mathematics that I want to study is more abstract than philosophy, and I believe it is far more important for the development of western civilization. Plato, the greatest philosopher of all time, would approve of my plans."

"Hrmmmmmpf!" his father grunted.

"I will depart for Geneva in the morning," Jacob continued. "I have made arrangements to tutor the children in the Waldkirch family there. One of the children, Elizabeth, is a girl who is blind. The father (a prominent businessman there) is convinced that all the children, including Elizabeth, are very bright. Since he wants me to teach Elizabeth to read and write and do arithmetic, in addition to teaching all the children such basic subjects as logic, physics, history, and all the rest, he needs a tutor who can be innovative enough to accomplish all that. He has learned that Girolamo Cardano (1501–1576)—a great mathematician in the last century—did some work on teaching a blind person to read and write.

"I have to admit that I had never thought before about whether it was possible for a blind person to learn to read and write, let alone how it might be accomplished. However, I have a description of Cardano's approach, and I'm hoping to improve on his methods. Although he was only partially successful in teaching his pupil how to read and write, I plan to do it right. I will succeed. I think this is an exciting project."

"I'm not impressed," his father muttered.

"Father, think about this a minute," Jacob said. "You want me to have a career in the Church, doing God's work on earth. You have to agree that teaching a blind girl to read and write is part of God's work also. Please give me some funds and the loan of a horse so that I may begin. After that I should be able to support my investigations in mathematics through tutoring. I must study with the great mathematicians of Europe. I will keep you posted on my whereabouts. Farewell, Father."

"Hrrmmmmmph," and Jacob's father left the room.

"Well," Jacob said to himself, "I guess the motto that I have chosen for myself fits: "*Invito patre sidera verso*—against my father's wishes I will study the stars." Jacob was comparing himself to Phaeton, the boy in Greek mythology who asked his father Helios, the sun god, to allow him to drive the chariot of the sun across the heavens for just one day. Although Phaeton's father had promised his son that he could have one wish, he never dreamed that his son would ask for this! It was a foolish wish, but the stubborn child reminded his father of his promise, and Helios felt impelled to keep his word. In the myth, since Phaeton was not strong enough to control the chariot of the sun—because unlike Helios he was not a god—the sun chariot was immediately in grave danger of crashing to the earth and destroying it. Zeus, the king of the gods, used his supernatural power and hurled a thunderbolt at Phaeton, killing him rather than allowing the rebellious boy to destroy the earth.

Like Phaeton, Jacob was sure that he could master his chosen chariot—astronomy and mathematics—but, unlike Phaeton, he would be able to reach for those stars in safety. There would be no need for Zeus or anyone else to interfere in his ambitious journey. Jacob was no fool, and his plan was something he knew he could carry out on his own. Jacob couldn't understand why his father refused to approve of the ideal life to which he was drawn—how could his father be so wrong? To Jacob, mathematics (and with it, astronomy) was the most beautiful subject imaginable, and he used his motto with relish for the rest of his life.

As he completed his studies, Jacob also chose a symbol to accompany his motto. It was the logarithmic (sometimes called equiangular)

Snail shell.

Jacob's seal, cloister of the Basel Münster.

spiral, which Jacob called the *spira mirabilis* [miraculous spiral]. As the size of the spiral grows (see picture), its shape remains the same. As the tangent follows the growing curve, the angle formed by the tangent and the curve's radial line remains constant. The chambered nautilus shell (or a snail's shell—see picture) is a famous example, formed by the shellfish as it grows larger and larger. Jacob wanted to have this spiral on his gravestone, although the actual spiral that appears there in the cloister of the Münster in Basel is only an approximation of it. Jacob's spiral is accompanied by the words in Latin, "*RESURGO EADEM MUTATA*" [Although changed, I shall arise again the same], as the curve does forever.

6
Jacob Travels to Geneva and Meets Elizabeth Waldkirch and Her Family

The three-day trip from Basel to Geneva took Jacob first through the Swiss towns of Biel and Neuchâtel, where he spent the night in a small inn, making arrangements for his horse to be well fed and well rested before the next long day on the road. The second day he traveled along the beautiful lake Neuchâtel and then on to the city of Lausanne. He was impressed with the vast lakes he found and with sailing boats skimming across the surface. From his childhood, he had known Basel's Rhine River with its powerful current. Although he had often crossed Basel's mighty Rhine in the small ferries that were powered only by the force of the river's current, and he had seen the large river boats that carried goods up and down the great river, these placid lakes were new to him. When he and his horse stopped along the shore of a lake to rest, Jacob dismounted and just gazed across the wide expanse of still water. Once, he even found people playing in the water, some of them apparently floating on its surface. Was that what people called swimming? Although he was a strong young man, he would never attempt to fight the powerful current of the Rhine River in his home city. He knew that he was no match for it! Perhaps it was different in a lake—the people that he saw swimming did not look as if they were any stronger than he was.

From Neuchâtel on, he found people who spoke only French, so it was a good thing he had spent some time working to improve

Rhine River at Basel.

his French before he set off on this trip. At Lausanne, Jacob caught his first glimpse of the snowy Alps. The dazzling Mont Blanc looked as if it were made of the purest salt! Here, he and his horse spent the night in another small inn before an early start on his final day of travel along Lake Geneva, bringing him by mid afternoon to the city of Geneva, the refuge of John Calvin, the founder of a dominant Evangelical Church in Switzerland. Everywhere he looked, there were spectacular mountains such as he had never even imagined. He rode his horse over the bridge that spanned the Rhone River at Geneva, and finally reached his destination.

"You must be *Monsieur* [Mister] Bernoulli," *Monsieur* Waldkirch greeted Jacob in French. "I am delighted that you were willing to come to Geneva to work with my children."

"Thank you so much for inviting me here!" Jacob exclaimed, also in French. "I have to admit that I have seen sights that I never dreamed of on this trip. I had no idea Geneva was such a beautiful city!"

"Yes," *Monsieur* Waldkirch said, "the Rhone River is nothing like your powerful Rhine, but our river and our lake have their charms. Did you know that after Lake Balaton in Hungary, our Lake Geneva is the largest lake in all of Europe? However, you didn't come for a lecture on the beauties of Geneva! Please allow me to begin by welcoming you to our home."

"Thank you so much, *Monsieur* Waldkirch," Jacob said. "I expect I will be learning much about your city during the time that I will be here. I should tell you that I am truly delighted to accept the challenge of teaching your children. I am particularly intrigued with the prospect of teaching Elizabeth. I expect we will all get along splendidly."

"Well, I hope you will have great success," *Monsieur* Waldkirch said. "I think you will find that Elizabeth is extremely bright. Are there any supplies that I need to arrange for you?"

"Yes, *Monsieur*, I'm afraid there are," Jacob said. "I will need to find a carpenter or wood carver who can make me models of the letters and numbers so that your blind daughter and I can begin to work. At first her learning will have to be exclusively tactile—by feel. Do I understand that she speaks German as well as French already?"

"Oh, yes," *Monsieur* Waldkirch said. "I'm pleased to say that she seems to have a real flair for languages. But, let me ask you if you would prefer to teach the children in German rather than in French."

"Oh, no," Jacob quickly replied. "I believe it will be best for them, particularly for Elizabeth, to learn at first in their native language, and I think my French is up to the task."

"Yes, *Monsieur*," *Monsieur* Waldkirch said, "your French is excellent. I agree that it would be preferable if you can teach her in French if you don't mind. I should tell you that Elizabeth has a superb memory, and that has always been a real advantage for her. We never need to tell her anything more than once.

"However, I should have thought to arrange for the wooden alphabet and numbers before you arrived. I'm so sorry! But I guess it's too late now. Now that you are here, perhaps you would like to make the arrangements yourself since you have a better idea of what you need than I would. My friend Simon Cartier is a carpenter and wood carver who lives on the road into the city, and I think you will find that he does excellent work. You must have passed his shop on your way today. Just tell him what you need and ask him to put the charges on my bill."

"Excellent," Jacob said. "The letters and numbers will need to be nicely finished, of course, so that your daughter can comfortably trace the shapes with her fingers. I don't want her to get a splinter in her finger! Shall I go talk with *Monsieur* Cartier this afternoon? I really cannot begin with *Mademoiselle* until I have those models."

"Of course," *Monsieur* Waldkirch said, "if you are sure you are not too tired from your journey. I'll ask the groom in the stables to provide you with a fresh horse (yours must be exhausted after three days of travel) and directions to find *Monsieur* Cartier's shop.

"*Monsieur* Cartier," Jacob began as he entered the wood-working shop, "my name is Bernoulli, and I will be tutoring *Monsieur* Waldkirch's daughter Elizabeth. He thought you would be able to make the supplies that I need."

"I'm so glad you are going to work with little Elizabeth! What a charming child!" Simon Cartier said. "I think you will find that she is a very clever pupil. I'm sure her father has told you that she is very bright, and that is no exaggeration. What would you like me to make for you?"

"What I need is a set of letters and numbers made of wood so that she can feel the shapes and can get to know the symbols," Jacob explained. "If possible, I'd like you to make two of each letter and number, each on its own rectangular block of wood, all the same size and about this thick [Jacob showed a length of about a half inch between his thumb and finger], with the letter or digit carved out on one face of the block so she can feel the shape. I would imagine it will be easier for you if you don't make them too small. However, if it is at all possible, I would like them to be small enough to fit into a cloth bag. The blocks will also need to be sanded very smoothly so that they are a pleasure to touch."

"This sounds very sensible to me," *Monsieur* Cartier said. "As a woodworker, I love the feel of a beautifully sanded piece of wood! You should realize that this will be a labor of love for me—I am very fond of Elizabeth."

"I'm so glad!" Jacob said. "In addition to the digits and letters, I will also need some open boxes, a size that will allow one digit or one letter to fit perfectly into each box. That way I will be able to teach *Mademoiselle* how to form larger numbers and words so that she can get the spacing right," Jacob said. "For arithmetic, I'll need a box for the ones' place, a box for the tens' place, a box for the hundreds' place, and so on. For words, I guess I'll need even more boxes, but I expect we'll be able to use the same boxes for both numbers and letters. Do you think you can make all of those?"

"I'm sure I can," *Monsieur* Cartier said. "How many boxes do you need, and how soon do you need all these things?"

"I think 30 boxes should be enough, because once she understands the spacing she should be able to move beyond the boxes," Jacob said. "I'm afraid I would like to have everything as soon as possible because I really cannot begin my work with her until I have them. Perhaps you could prepare one set of the numbers and a few of the boxes first, so that we can get started on arithmetic. Then you could complete the rest of the sets while I'm working with her on the numbers. I imagine it will take her awhile to learn them."

"Would Monday be soon enough for the numbers and the first boxes?" Simon asked.

"Yes, Monday will be fine," Jacob said.

"By the way, do you need both capital letters and lowercase letters?" *Monsieur* Cartier asked.

"Yes, I will need both," Jacob said, "but I think one set of capital letters will be enough. However, I think there will not be such a rush on the letters. I suspect arithmetic will be a real challenge for her."

"*Monsieur* Bernoulli, I think you will be surprised at how quickly she learns," *Monsieur* Cartier said. "She is an unusually intelligent girl. I'll have my man deliver one set of numbers and several boxes to you at the Waldkirchs' home on Monday morning, and I'll try to have the letters and the rest of the numbers as well as the rest of the boxes ready by the end of the week. I like this project very much. I assume I should put this on *Monsieur* Waldkirch's bill."

"That's what *Monsieur* Waldkirch asked us to do," Jacob said.

"Good," *Monsieur* Cartier said. "Shall I ask my wife to make a bag for the pieces?"

"That would be wonderful, *Monsieur* Cartier! Thank you!" Jacob said as he remounted the borrowed horse and set off once again for the Waldkirchs' home.

When Jacob joined the family for supper that evening, he met all the children as well as their mother, *Madame* [Mrs.] Waldkirch, for the first time. The atmosphere in the home was warm, and Jacob was impressed with how poised Elizabeth was. He learned that she had lost the sight in both eyes because of an infection just two weeks after she was born. This meant that she could never remember seeing anything. However, she handled the dishes on the table easily, never spilling anything. All the children were articulate, carrying on a conversation in both French and German with no trouble. In fact, Jacob had to admit that Elizabeth's German was at least as good as his French. When he commented on this, her father explained that she could also speak Latin. Jacob decided that his assignment with this very bright child was decidedly possible, and he found that he liked the Waldkirch family very much. The family seemed happy, with lots of good fun as well as serious talk during the meal.

7
Jacob Teaches Elizabeth Waldkirch to Read and Write Numbers and Words

When the digits and boxes arrived on Monday morning, Jacob was delighted with them. All the surfaces were beautifully smooth, all the edges and corners had been expertly rounded off, and the drawstring bag was beautifully finished as well. Jacob began to work with Elizabeth at once. He gave her the digits one at a time, encouraging her to handle them for long enough to learn their shapes well. Fortunately, she already knew how to count and do simple arithmetic in her head.

He urged her to be patient at first, but he soon realized that Elizabeth had learned about patience from an early age. Jacob was the one who needed to be reminded about patience. This was his first experience as a teacher other than his informal sessions with his brother Johann. He was determined to succeed, but he needed to remind himself repeatedly that what was obvious to him wasn't necessarily obvious to her.

"*Mademoiselle*," Jacob said, "First, you will need to learn to recognize the shapes of all the digits. Please note that the digit 1 is a straight line with just a little hook at the top. Can you feel that?"

"Yes, *Monsieur*," Elizabeth said.

"Now I want you to feel the digit 2. It has a straight line across the bottom, but then it curves from the left end of the base up to the right and then around to the left, making a graceful loop. Do you feel that? Wait a minute! You do know your left from your right, don't you?"

"Yes, *Monsieur* Bernoulli," Elizabeth said. "This is my right hand. But excuse me for asking, *Monsieur.* What do you mean by digits? Is digit just another word for number?"

"No, *Mademoiselle,*" Jacob said. "There is an important difference. The digits are the symbols that we use to write the numbers. We have ten digits—0, 1, 2, 3, 4, 5, 6, 7, 8, and 9. They are the symbols that you are learning now. I'll teach you how we use the digits to construct numbers as soon as you know the digits."

"But my father never used the word digit with me," Elizabeth protested.

"That's because he was not teaching you to read or write," Jacob explained. "If you are going to read and write numbers, you must begin with the digits. Until you learned to read and write numbers, however, there was no need to distinguish between digits and numbers."

"Thank you, *Monsieur,*" Elizabeth said. "I want to learn all of this well and quickly."

"Good for you!" Jacob said.

"Thank you, *Monsieur,*" Elizabeth said, "but can you tell me this: Is the digit nine just the upside down version of six? Can I make a nine by simply turning the six upside down?"

"Yes, I suppose you could," Jacob said, surprised at that obvious fact that he had never considered seriously before. "Can you see how both 6 and 9 curl around into themselves? They really form a spiral, a shape that I find very appealing."

"Yes, *Monsieur,*" Elizabeth said, "I like that too, but now could you please tell me about zero? It seems to be round with a hole in the middle. My father told me about zero, but I have never been able to understand why we need it."

"Actually, it's not completely round. It is really somewhat longer from top to bottom than it is from left to right. Please take this zero in your hand. Can you feel that difference? " Jacob asked. "It's what we call an oval. However, I should also answer your excellent question about the meaning of zero. The number zero simply means that we have none of the thing at all. If you have zero dolls, that means you don't have any, but you already knew that!"

"Yes, *Monsieur*, I know what it means to have no dolls or no bread, but why do we need a symbol for it? Why do we bother to count something if it isn't there?" Elizabeth asked.

"Sometimes we need to explain that a container or a group is empty, and the number zero is useful for that," Jacob explained, "but the digit zero is really far more useful than the number zero. However, I'm getting ahead of myself. I'll get to that a bit later."

"Of course, *Monsieur*. So you are telling me that zero does more than tell us that we don't have something. Thank you, *Monsieur*," Elizabeth said. "That's something I had been wondering about. I can wait for you to tell me more about it later, but please don't forget."

"Don't worry about that, *Mademoiselle*!" Jacob said. "I consider the digit zero extremely important. Now, let's review the digits one more time. What is this digit?" and he handed her a block.

"That is five, isn't it?" Elizabeth asked.

"That is correct, *Mademoiselle*," Jacob said. "Now, how about this digit? No, *Mademoiselle*, you must hold it right side up. It does make a difference."

"Yes, *Monsieur*. I'll try to be more careful," Elizabeth said. Then she reached over to where Jacob had placed the rest of the digits on the table—she knew exactly where they were—and she named each digit correctly as she picked it up, this time being careful to hold each digit right side up.

"*Mademoiselle*, I believe you know all ten digits now," Jacob said.

"I think so," Elizabeth said, "and I like them."

"Good. So now we can move on to the construction of numbers," Jacob continued, "and this is where the distinction between digits and numbers is important. You see, our number system uses place value—the location of a digit in combination with the value of the place in the number—which go together to tell us what a number is worth. The word for digit in French [*chiffre*] comes from the Arabic language. We have the Arabs to thank for our number system, so it is appropriate for us to use the Arabic word for the symbols. You might be interested that the German word for digit also comes from the same Arabic root."

"So, were the Arabs the first people to write numbers, *Monsieur*?" Elizabeth asked.

"No, the ancient Sumerians and ancient Egyptians wrote what I would have to describe as primitive number symbols using a combination of dots and lines many centuries before the Arabs," Jacob said. "The ancient Greeks and Romans wrote numbers also, but they used letters from their alphabets, and they also did not use place value in the modern sense. Greek and Roman written numbers were very awkward and often involve many symbols. Arabic numbers are vastly superior. We are very fortunate to have our number system. It makes calculating easy."

"That is very interesting," Elizabeth said. "Perhaps sometime you could tell me how the Greeks and Romans wrote numbers."

"I could do that, but I think we need to work with our own number system first. We work from the right as we construct a number," Jacob said. "The place on the right is the ones' place. We will indicate that by this first box. Please touch it with your hand, *Mademoiselle*. The next place, just to the left of the ones' place, is the tens' place. A digit in the tens' place is worth the value of the digit multiplied times ten in exactly the same way that a digit in the ones' place is worth its name times one. Does that make sense to you?"

"Yes, *Monsieur*," Elizabeth said. "Does that mean that the third box is the hundreds' box?"

"That's right," Jacob said. "What do you think a three would be worth in the hundreds' box?"

"It would have to be 300, wouldn't it?" Elizabeth asked.

"That's right!" he said. "Now the amazing fact about our number system is that we can write any number, no matter how big or small it is, using only these ten digits and however many boxes we need."

"Is this where we come to the use of the digit zero?" Elizabeth asked, unable to control her curiosity any longer.

"That's right, *Mademoiselle*," Jacob said. "What we need the digit zero for is to indicate that a box is empty. Since most people write numbers without boxes, we need a symbol to show that a given place is empty. So if there is a zero in the ones' place and a five in the tens' place, that means we have five tens and zero ones, so that number would be the number 50. Does that make sense?"

"Yes it does, *Monsieur*," Elizabeth said. "Would we write the number 500 by putting a five in the hundreds' place and then zeroes in the tens' and ones' places?"

"That's right, *Mademoiselle*," Jacob said. "Shall we try another number now?"

"Yes, please!" Elizabeth said.

"Okay, I have put a digit in each of these boxes," Jacob said. "Remember that the box on the right is the ones' box, the box in the middle is the tens' box, and the box on the left is the hundreds' box."

"Where should I start?" Elizabeth asked. "Should I start on the right?"

"Yes, let's do that for now, although after you have identified the parts of the number, we will actually read the entire number from the left," Jacob said.

He was delighted as he saw how quickly she figured the numbers out, correctly reading four- and five-digit numbers within only a few minutes. Before the end of the week, she was doing serious arithmetic with her boxes of numbers and even writing them on paper with a piece of charcoal. Jacob was pleased to see that her numbers were perfectly clear to any seeing reader. Filled with excitement, Elizabeth quickly took the paper and ran to show her mother what she had done. It was an amazing accomplishment! Her mother was quick to tell her that she knew her father would also be thrilled. After *Monsieur* Cartier's man delivered the letters, reading and writing proceeded just as quickly, and within several months both Jacob and Elizabeth were delighted. She truly could read and write.

One morning after they had worked for several hours together, Jacob said to Elizabeth: "*Mademoiselle*, I have a question for you that has nothing to do with reading and writing, but it is something I have been wondering about. I'm afraid it is rather personal. I hope you don't mind."

"Of course not," Elizabeth said. "You have answered all my questions, so there is no reason for me not to answer yours. What do you want me to tell you?"

"Thank you. I'm curious about how you dream when you are asleep," Jacob said. "When I dream, I see things in my mind. Can you tell me what your dreams are like?"

"Oh!" Elizabeth said. "I never thought about that. I guess I don't see things in my dreams the way you probably do, but things do happen. In fact, I think my dreams have been changing since you have been teaching me. I sometimes find myself handling the shapes of the letters and digits in my mind as I dream. It is almost as if I was awake and using them!"

"That is very exciting, Elizabeth!" Jacob said. "That means that in fact you are seeing. Thank you so much for telling me about that."

Because the children could not be expected to spend all their time on their lessons, Jacob had the occasional afternoon to himself. Some days, after a long morning of tutoring, he took his horse into town so that he could introduce himself to scholars at the academy in town and find out what mathematics resources were available. He was pleased to get to know several instructors and students in mathematics. However, although some of them seemed to be interested in mathematics, Jacob soon realized that he was far ahead of them all.

Explaining the *Coss* and Pappus to his new colleagues was challenging, but Jacob realized that it was helping him, too. In the process of explaining the subject, he was coming to understand it at a deeper level. To please his father, Jacob also engaged in several debates with theologians in the city and even delivered a sermon in one of the churches in town.

During his 20 months in Geneva, Jacob also had the chance to study Cardano's *Ars Magna* [*Great Art*, or the rules of algebra], which he was able to borrow from a friend of his employer in Geneva. It was this gentleman who had told *Monsieur* Waldkirch about Cardano, the mathematician who had first attempted to teach a blind person to read and write. Although Cardano's mathematics book had been published in Basel in 1570, Jacob had not been able to find it in the Basel University library.

During his time in Geneva, Jacob also developed some skill at the game of tennis. While still at home in Basel he had occasionally picked up a racket, but it was only in Geneva that he was able to play tennis regularly and develop his physical coordination for this sport. Since the local sport club had fine tennis courts, Jacob had many occasions to play. In later years, he explored probability as it relates to games such as tennis.

8
Sundials, and Tutoring in France

At the end of his time in Geneva in the spring of 1677, Jacob, who was then 23 years old, received word of a position in France tutoring the children of the Marquis de Lostanges. The marquis offered to pay for his journey by carriage to Nède near Limousin in south-central France, where once again he would serve as a tutor. Since Jacob had determined that he needed to journey to France to continue his study of mathematics, he was pleased to take advantage of this offer. Once again he found time to explore mathematics during some free afternoons. At this time, Jacob made a serious study of sundials and their construction, perfecting a method for determining the angle for the gnomon (the rod that creates the shadow from which the time is read) to accommodate the latitude of a specific location based on the tilt of the earth at that point. The angle is critical if the sundial is to be usable during daylight hours throughout the year. A properly constructed sundial's only limitation is the need for clear sunny weather and a southern exposure.

"*Monsieur* Bernoulli," the marquis asked him one afternoon, "Could you tell me about that table you have there?"

"Yes, *Monsieur*," Jacob replied. "This is a table showing the angle at which the gnomon of a sundial must be mounted at any given latitude in order to construct a reliable sundial. It is based on the inclination of the earth with respect to the sun. I have read about sundial construction, and I was curious to see if I could generalize the technique. It shouldn't be necessary to start from scratch with the calculations any time we want to place a sundial in a garden."

"Fascinating!" the marquis said. "And where did you find the table, or did you make the table yourself?"

"Yes, *Monsieur*," Jacob said, "as far as I know there was no such table available, so I made it myself. I have been doing the calculations for the table during my free moments over the past few weeks, using astronomical data to find the exact angle required at each of the latitudes. Apparently, no one else has taken the time to do that. Depending on the latitude of a location, I can easily get the angle just right, so that using my table I can construct a reliable sundial to be placed anywhere in France or the Helvetian Confederation."

"So, what would be the correct angle for a sundial here in Nède?"

"Here it is," Jacob said as he quickly scanned the table and sketched the angle.

"Would it be possible for you make a sundial that could be placed in my garden here in Nède?" his employer asked. "I would be delighted to pay you for it."

"Well, it would have to be in a location that gets sun throughout the day, possibly a wall with a southern exposure or an open spot in the middle of the garden away from large trees and buildings that might cast inconvenient shadows," Jacob said. "I must say, it sounds like an interesting challenge. I'd be glad to do it. I would love to put my table to a practical test."

"Then come over this way, please," his employer said. "I think I have a wall in my garden that would be just right." Suddenly the marquis noticed that Jacob was walking very slowly—he was not able to keep up with him. "Are you coming, *Monsieur* Bernoulli?"

"Yes, *Monsieur*, I am coming," Jacob said, "but I'm afraid I can't walk too quickly. Ouch! My toe!"

"I'm so sorry, *Monsieur*," the marquis said. "I can tell that you are in terrible pain. Perhaps we should talk about the sundial another time."

"Oh no, I'm fine," Jacob said, hobbling as quickly as he could and trying not to show his distress. At this time Jacob was beginning to suffer from serious health problems, often resulting in difficulty walking. Lately he was finding the game of tennis entirely too painful. He had been soaking his foot morning and evening all week, but still his big toe was badly inflamed. For the rest of his life, he suffered from severe pain in his legs and feet, probably the result of gout and perhaps from scurvy (a deficiency of vitamin C) as well. At that time, the standard European diet during the winter included very few fruits and vegetables, the natural sources of that vitamin. Unfortunately, Jacob's pain sometimes distracted him from the scientific research he wanted to pursue, and eventually his illnesses would cut short his life.

"*Monsieur*, is this the wall you were talking about?" Jacob asked stoically.

"Yes," the marquis said. "What do you think?"

"I think it would be perfect," Jacob said, holding his arm at the approximate angle so that he could see its shadow.

"That will be wonderful!" the marquis said enthusiastically. "I have wanted a sundial here in my garden for several years, but until today I didn't see how I could get one. You are a very clever young man!"

"I'll need to have a blacksmith make a straight rod for me," Jacob continued, "but that should not be difficult for him. Do you have a mason who could implant the rod at the angle that I specify so that I can proceed from there and place the markings on the wall?"

"Of course! How long should the rod be?" the Marquis asked.

"I would think about as long as the distance from a man's elbow to the tip of his finger," Jacob said. "I wouldn't be surprised if the smith even has such a rod on hand."

"I'll instruct the smith to have one ready for me by early tomorrow morning," he said. "My mason is coming tomorrow afternoon. I'll send word to them both to make those preparations."

"Then the rest should be no problem," Jacob said. "I should be able to make my plans this evening."

"Good. I'll send a messenger to the blacksmith now," the marquis said, "and then I'll send a message to the mason to pick up the rod on his way to my house tomorrow."

"That should be fine," Jacob said. "I will also need some paint and a fine paintbrush so that I can do the markings on the wall. Shall I write something on the dial, perhaps *tempus fugit* [Latin for 'time flies']?"

"That would be good!" the marquis said enthusiastically, "or how about '*sic vita fugit*' [thus life flies] for a little irony? You are too young for such thoughts, but by the time you are my age, you will begin to think about the passage of the years as well as the hours!"

Jacob's sundial on the wall of the marquis' house was a charming work of science and art, which allowed the family to know the precise time on any sunny day. Visitors to the garden were always impressed (the marquis was delighted to demonstrate that he possessed the latest technology), and one neighbor was so impressed that he asked Jacob to make him a sundial as well. Chuckling at the message on the marquis' sundial, he suggested that his sundial's message could be the simple "*Je ne compte que les heures ensoleillées*" [I count only the sunny hours]. Jacob was delighted.

The neighbor chose an open spot in his garden for which Jacob designed a handsome sundial to be constructed on top of a large flat stone. Once again, Jacob needed the help of a stone mason and a blacksmith to erect the flat stone in the garden and to plant the gnomon in the stone at the precise angle that Jacob gave him. Once it, was mounted, Jacob made the markings on the stone to designate the hours and then wrote the saying. Jacob was developing a reputation in the neighborhood not only as a savant, but also as a clever but practical scientist.

Once again keeping his word to his father, Jacob preached several sermons in area churches around Nède. By this time, Jacob's French was excellent and his sermons were well received. Although

Jacob had already decided that a life in the ministry was not his first choice, he was enough of a realist to keep his options open. He had made a serious promise to his father which, as an honorable young man, he intended to keep. This was wise since he had not yet met anyone who had actually made a career in mathematics.

At this time Jacob began his *Meditationes*, his scientific diary, in which he recorded his explorations into mathematics and physics. He wrote it in Latin—technically a "dead language" for more than 1,000 years, but still the living language of scientists throughout Europe. As a well-educated young man, Jacob wrote easily in Latin, often even writing poetry in Latin for his own and his friends' amusement.

In 1679, Jacob journeyed farther west to the city of Bordeaux, to tutor the son of a local official there—a notary who was certified to prepare and sign off on official documents. At this time Jacob was accumulating as much money as he could from tutoring and consulting about sundials, meeting with scientists wherever he could find them, and reading anything he could find on mathematics.

Jacob was fascinated to observe firsthand the phenomenon of tides as he walked around the seaport of Bordeaux. He had learned about high tide and low tide as a school boy in his studies of geography, but as he watched the water level actually rising and falling in the harbor, Jacob was astonished. The Rhine River in Basel certainly did not behave this way! The tides, which at first had struck him as unpredictable, were something the men who worked in the harbor dealt with every day. Their understanding, however, was only at a practical level. Jacob wanted to understand why the tides happened, and he wondered how precisely they could be predicted. Jacob was an ambitious young man.

After completing his duties in Bordeaux, Jacob had accumulated enough money that he could then travel to Paris and study full time. His first goal was to read the works of Descartes, who Jacob knew was the most esteemed mathematician of the current century—probably the most important French mathematician ever. He had been trying everywhere to find someone who had a copy of Descartes' work *La Géométrie*, which had been published in French in 1637 and later in Latin in 1649 and 1659—translated by the Dutch mathematician Frans van Schooten (1615–1660). Jacob still had not found a copy of this work in either French or Latin.

9
Jacob Meets with Mathematicians in Paris

In 1680 at the age of 26, Jacob traveled to Paris, where he arranged to meet with several philosophers and mathematicians. He was particularly impressed with Nicolas Malebranche (1638–1715), a nobleman who was a priest and a philosopher and whose library included the works of François Viète (1540–1603) and René Descartes (1596–1650). Malebranche had spent much of his life studying Descartes, concentrating on both his philosophic works and his mathematics. Malebranche was the kind of scholar Jacob had been hoping to meet.

"*Monsieur*," Jacob said, "I have heard of Descartes, whose work I have been eager to read for some time, but all that I know of Viète is just his name. Did you say that he came before Descartes? Do you think that it is important that I start with the mathematical writings of Viète or could I simply begin with Descartes?"

"Well, *Monsieur*," Malebranche replied, "there is no denying that Viète was important. He did his major work about 50 years before Descartes, and I think it is clear that Descartes learned from Viète's work, although I don't know that he ever admitted any debt to him. There is one good story that you should hear about Viète if you know nothing about him.

"In 1590, during a war with Spain, Henri IV, the king of France, obtained some intercepted letters from Spain written to nobles in the French court. These letters were written in cipher, and the king was determined to find out what they said, since he thought they might have information that would be valuable to him. If they didn't, the king asked, then why would anyone have bothered to write them in cipher? The cipher was difficult, and no one among Henri's advisors could even begin to decipher them.

"However, in his court Henri also had a mathematician (that would be Viète), so he asked Viète to try to decipher the letters. After a few months' effort—it was certainly a difficult code!—Viète cracked the cipher, and Henri was able to foil the Spanish king. The Spanish king, for his part, couldn't believe that anyone could decipher his message, and he immediately accused Henri of using witchcraft! The members of the Spanish court had sworn that no one, anywhere—particularly not a bungling Frenchman (!)—would ever be able to decipher it. I love that kind of witchcraft, don't you?" Malebranche asked Jacob.

"That is a wonderful story!" Jacob said. "I wonder if it would impress my father, who has fought against my studies of mathematics since I began studying at the university."

"I'm sorry he's done that," Malebranche said. "I hope you won't ever let him keep you from it. I cannot imagine life without mathematics."

"Well, I'm here, and I'm learning mathematics!" Jacob said. "You may be sure that I will study mathematics regardless of what my father says."

"That's good," Malebranche said. "Getting back to your question, I suppose it is reasonable to say that there is no need for you to read Viète's work yourself, although it would be unfair not to give him credit for inspiring some of Descartes' important work, including the use of letters to represent known and unknown quantities. Viète's plan was to denote unknown quantities by vowels (*A, E, I, O*, and *U*) and known quantities by consonants (*B, C, D, F*, etc.), while Descartes chose instead to use the letters at the end of the alphabet to stand for unknown quantities (x, y, and z) and letters at the beginning of the alphabet to stand for known quantities (a, b, and *c*). I, personally, doubt that the choice of letters is significant, although Viète was limited to five unknowns since there are only five vowels, and I suppose that could possibly pose a problem sometime. We truly have Viète and Descartes to thank for what I like to call literal algebra—algebra using letters—which I believe will soon be all anyone will ever use for algebra. The *cossists* are already folding up their tents—they know they have already lost. You will see. Literal algebra is truly the mathematics of the future."

"I can hardly wait to get started!" Jacob said. "I have to admit that I have always found Rudolph's words and abbreviations cumbersome."

"You are right about that," Malebranche continued. "Another difference between Descartes' and Viète's algebras is that Descartes used a superscript—an exponent—when he wanted to indicate $x \cdot x$, writing it as x^2 or $y \cdot y \cdot y$ as y^3, and I believe that may be significant."

"So the exponent tells how many times the quantity is multiplied times itself? I like that!" Jacob said. "Rudolph could have used that in his *Coss*!"

"Yes," Malebranche said, "and Viète only 50 years earlier still used only verbal or syncopated symbols such as "*A quadratum*" [*quadratum* is Latin for squared] or "*A* quad" (in much the same way that Rudolph did in the *Coss*), whereas Descartes wrote as it as a^2. I suspect Descartes' notation will be the one that survives, but we'll have to wait and see. Descartes' work is certainly much better known than Viète's today, and probably with good reason. I believe we are working at a very exciting time in the development of mathematics. Do you suppose someone 200 or 300 years from now will simply consider Descartes' work the norm?"

"That is possible," Jacob said. "But I have to admit that I have struggled simply to find a copy of Descartes' work for most of a year, so it still isn't as easy to find as it should be."

"No, it's a pity," Malebranche agreed.

Jacob continued, "I think the exponent—is that the word you used?—sounds like an excellent idea as a substitute for *quad*, but it may take me awhile to get used to it. I would think it would be faster, and certainly it's a pity to slow down our mathematics just because of inconvenient notation. *Coss*' abbreviations were an improvement over the verbal mathematics of the classical mathematicians."

"Would you like to hear an interesting little tidbit about Descartes' use of letters for variables?" Malebranche asked.

"I would love to!" Jacob said. "You are a gold mine of information on mathematics!"

Malebranche continued, "I have read that Descartes planned to use the letters x, y, and z to stand for his unknown quantities, and he hoped mathematicians would use a variety of those letters. However, he lost on that point. Apparently his printer had some difficulty with the availability of letters. He found that he was running low on his supplies of y and z. As you know, the French language uses those two letters a great deal, but it uses the letter x much less often. So the result is that the printed version of Descartes *Géométrie* uses x as a variable most of the time. It was a practical solution to a practical problem, having no mathematical significance at all. The Latin translation of Descartes' work has continued that, even though the printer's problem does not arise in Latin. I was amazed when I read this. I wonder if mathematicians will continue to use mainly x for the variable in the future. That would not have pleased Descartes, may he rest in peace!"

"That is very interesting!" Jacob said. "I wouldn't have expected it to be a practical issue, but I can see the printer's problem. Now in German, we don't use the letter y anywhere near as much as French does, and we probably use x even less than the French, so a German publisher, given a choice, might have been willing to alternate the letters x and y, satisfying Descartes at least in part. Interesting! But wait! Why didn't Descartes write in Latin? That is the language of science."

"I don't know why, but he didn't," Malebranche said. "He wrote in French, although now his work has been translated into Latin so that mathematicians throughout Europe can read it where it is available. You have unfortunately found that availability is a serious problem. I'm sure Descartes knew Latin, and he was living in Holland at the time, so I can't explain it."

Malebranche provided Jacob with a copy of Descartes' *La Géométrie* in French, the original language, and Jacob opened it and began to study immediately. Although by this time Jacob's French was perfectly fluent, this was difficult reading. However, Jacob knew what he wanted, and he knew he was smart enough to master it. He read with quill, ink, and paper, working actively as he had learned to do with Pappus' works a few years earlier—Jacob knew that anyone who reads mathematics without quill and paper is not really serious about understanding it.

"*Monsieur* Bernoulli," Malebranche said the next afternoon, "you are probably finding Descartes difficult to read."

"Yes, it is difficult," Jacob answered, "but I think I can do it. Did you have trouble figuring it out for yourself?"

"Yes, I encountered the occasional road block, but through working at it seriously day after day and talking with other mathematicians here in Paris, I managed," Malebranche said.

"Well, then," Jacob said, "I guess I should be able to do that too."

"As you encounter difficulties, don't hesitate to ask me for help," Malebranche said. "I would hate to have you waste too much time on the basic concepts. That might not leave you enough time for the more interesting parts."

"Thank you, *Monsieur*," Jacob said. "So far I am doing all right."

"Did you know that Descartes deliberately made it difficult to read?" Malebranche asked.

"I wondered about that," Jacob said. "Do you know why he did it?"

"Well, I understand that he justified it in a couple of ways," Malebranche said. "First he said that he had given enough information so that anyone who had the proper background could figure it out. He thought that any more information would simply be redundant. He didn't want to insult his readers."

"I can accept that," Jacob said.

"Then," Malebranche continued, "Descartes said that he wanted his readers to have the genuine pleasure of completing his arguments. If we view mathematics as a sport, it would be inconsiderate of a mathematician to give it all away immediately. Mathematics is truly a treasure hunt—if someone tells you before you start where the treasure is hidden, it is no fun at all."

"You know, *Monsieur* Malebranche, that makes sense," Jacob said. "I guess I respect Descartes more after hearing that."

"Yes, but if what you want is to understand the mathematics so that you can pursue his ideas further, it would be futile to waste too much time on the foundations," Malebranche said. "I believe you are very serious about moving along in your studies of mathematics."

"Yes, indeed," Jacob said.

A little while later, Malebranche was sitting, looking at Bernoulli. Finally, he said, "Excuse me, *Monsieur* Bernoulli, wouldn't you be more comfortable using this footstool so that you can elevate your foot. It appears to be causing you serious pain."

"Well, *Monsieur* Malebranche, I appreciate the offer," Jacob said, "but my foot isn't bothering me too much. I don't want to impose. And besides, I believe you need the footstool more than I do." Malebranche had suffered from birth with a severe curvature of the spine, causing him persistent pain and limiting his mobility all his life. Jacob and Malebranche hadn't discussed it before, but Malebranche's suffering was obvious.

"I will ask the servant to bring us a second footstool," Malebranche said. "I have several. We are two diligent scholars who need whatever devices are available to help us in our pursuit of knowledge. Our study should not be hampered by physical pain any more than necessary."

Jacob struggled through *La Géométrie*, drawing sketches as needed, and finally comprehending the entire work, for the first time seeing algebra as the best way to study geometry. By this time he was adept at using an exponent to show a power of a variable and *x* (if not *y* and *z*!) for his unknowns. Descartes had thought it through carefully, producing brand new mathematics—what is sometimes now called analytic geometry—out of his own imagination. Jacob correctly saw it as the work of a genius.

Jacob and Malebranche also spent time discussing Descartes' philosophy—a topic that interested Malebranche (who was a priest in the Roman Catholic Church) far more than Jacob at this point. However, since Jacob recognized his debt to Malebranche in making the *Géométrie* available to him, he joined these discussions with apparent enthusiasm. His university studies in philosophy and theology had prepared him well for such debates. The two men discussed Descartes' famous statement *cogito ergo sum* [Latin for "I think, therefore I am"], and the difficulty of rationalizing Cartesian philosophy with the theology of the Church of Rome. Malebranche was convinced that Descartes' philosophy could be adapted to the teachings of the Catholic Church, even in the dispute over transubstantiation, although many Roman Catholic theologians found the Cartesians' approach too close to that of the hated Protestants'. Is the bread that is used in the Eucharist actually transformed into the body of Christ (the Roman Catholic view) or is it only a symbol of the body of Christ (the Protestant view)? A more fundamental question explores the relation between faith and reason.

While he was in Paris, Jacob also did some work on astronomy, another subject that his father had prohibited him from studying. The Latin motto that he had taken for himself—against my father's wishes I will study the stars—was true. He studied them in earnest. Using a borrowed telescope, he studied carefully the path of a comet in 1680. He concluded that a comet is not ephemeral—it doesn't appear for a brief time and then evaporate, as was the common belief at that time—and that a comet travels on a predictable path, orbiting the sun in the same way that the planets do, although often in a much larger orbit. His calculations convinced him that the 1680 comet should return on 17 May 1719. Whether or not that prediction was true (and his calculations were not correct as it turned out), then it is ludicrous to say that a comet is an omen of some calamity. Despite Jacob's mistake, he was correct about comets in general. A comet is not a fleeting sign from heaven indicating imminent misfortune. Therefore, Jacob said, it was foolish for people to make decisions based on that false reading of the heavens.

As he talked with people in Paris, however, he found that his radical view was not popular, so he decided to adjust it slightly. He then wrote that the head of the comet is not an omen—it cannot have anything to do with future events here on earth—but that he couldn't be absolutely certain that the tail does not indicate something. He announced that the tail is changeable and thus its shape might possibly have some significance. That appeased his critics without opening him up for criticism from scientists. It was a mild concession that protected him from attacks from all sides.

10
Jacob Travels to Holland and England

In 1681, Jacob traveled to Amsterdam, where he may have met with Jan Hudde (1628–1704), the foremost mathematician in Holland—in fact, the most important mathematician in all of Europe at the time. Hudde was a serious scholar of Descartes' mathematics and, using Descartes as his starting point, Hudde had devised two rules for dealing with polynomial equations that moved mathematics further toward the development of the calculus.

Hudde, who had worked extensively with his teacher Frans van Schooten, the translator and editor of the expanded version of Descartes' geometry, was a logical person for Jacob to meet. It was in Holland—not France—that serious mathematics was being pursued at the time.

"*Monsieur* Hudde," Jacob began. "No, I'm so sorry, Sir. In France I was careful to address people in French. In Holland I would like to use Dutch, but unfortunately I don't know the Dutch language. How should I address you, Sir?"

"The Dutch equivalent of *Monsieur* is *Meneer* (Mr.), but it doesn't matter," Hudde replied. "Perhaps it would be easier if we simply communicate in French, which I believe we both speak easily."

"No, no!" Jacob said in French. "At the very least I would like to address you correctly, *Meneer* Hudde. Was that right?"

"That was fine, *Herr* Bernoulli," Hudde said to Jacob in German. "German is your native language, isn't it?"

"That's right, although it is really not important," Jacob said. "Allow me please to begin again, *Meneer* Hudde. I notice that you boldly use a letter as a variable to represent any real number when you write mathematics, regardless of whether it stands for a positive or a negative quantity. Descartes didn't recognize negative numbers, as I remember. Isn't it risky to allow the variable to stand for a negative?"

"But it is essential," Hudde replied. "You see, Descartes, brilliant though he was, ignored negative numbers. Nonetheless, they are legitimate numbers. If algebra is to help us, we certainly need to be able to represent negative quantities with variables. Otherwise we lose at least half of the value of algebra. You have to admit that a debt is just as real as a credit in the world of business, and that is just one small illustration of negative quantities in mathematics."

"Yes, I suppose that is true," Jacob said.

"Allowing the variable to stand for both negative and positive quantities has not interfered with my work in the least, *Herr* Bernoulli," Hudde said, "and it has helped me dramatically. Furthermore, when we are solving an equation in algebra we frequently don't know whether a quantity will end up being positive or negative until we reach a solution (sometimes it even turns out to be positive sometimes and negative at other times!), so clearly the variable needs to cover both signs. Take a look at this." As Hudde showed Jacob his latest work, Jacob could see that the variables for negatives were indeed allowing him to do some important work.

"Do you mind if I read this through, *Meneer* Hudde?" Jacob asked, indicating the work in his hand which was written in Latin, a language they both could read and write easily.

"If you want to do that, *Herr* Bernoulli, that is not where you should start," Hudde gently corrected him. Walking to a table in the corner of the room and picking up another essay, he continued, "I would recommend that you begin with this essay that I wrote a year ago. Otherwise my more recent work will not be as clear as you would like. You need to follow my reasoning in the order that

I wrote it. Please feel free to sit down here and read it. Can I offer you a cup of tea?"

"That would be delightful, *Meneer* Hudde!" Jacob said. "Thank you so much!"

From Amsterdam, Jacob went on to the town of Leyden where he stayed for ten months, getting to know the mathematics professors there and perhaps teaching several classes for them. Since he would have lectured in Latin, his Dutch students would have been able to understand him perfectly.

From Holland, Jacob went on to London, where he was eager to meet John Flamsteed (1646–1719), the Astronomer Royal, who would soon move into and direct the new Royal Greenwich Observatory, in a position that Flamsteed would hold for the rest of his life. Jacob also met with Robert Boyle (1627–1691), familiarizing himself with that scientist's brilliant work in chemistry. Jacob learned how Boyle had discovered the fudamental difference between mixtures and compounds in chemistry, and he listened carefully to Boyle's description of his research into the chemistry of combustion and the process of respiration in animals, a subject that Jacob's nephew Daniel would study in his own doctoral dissertation 40 years later..

Jacob also talked with Robert Hooke (1635–1703), looking with fascination at his beautifully illustrated volume *Micrographia*, showing the world of things so small that they could not be seen with the naked eye. Hooke also described to Jacob an exciting new plan for a tubeless telescope, whose eyepiece was mounted separately from the lens so that the distance between the two could be changed as needed. When Jacob pressed him, Hooke admitted that the first tubeless telescope was actually not his own invention and that he hadn't yet constructed one himself. The first one had been made by the Italian lens maker Giuseppe Campani (1635–1715), but Hooke was eager to construct his own. As a superb contriver of things both mechanical and optical, this was well within Hooke's abilities.

In London Jacob learned of the mathematical writings of the English mathematicians John Wallis (1616–1703) and Isaac Barrow (1630–1677). In his 1669 textbook on geometry, Barrow had included information on the new work on maxima and minima—finding the greatest and the least possible value for an algebraic expression—and a useful technique for finding them. Barrow did not claim that this was his own original work, but his explanations were clear, involving the construction of the tangent to a curve (the straight line that hits the curve at only one point and that demonstrates the slope of the curve at that particular point), and Jacob studied that too. A few years later he would realize that Barrow's geometry is actually part of the foundation of the developing field that would later be known as the calculus.

By the time Jacob returned to Basel in 1681, he had mastered both Barrow's and Wallis' work, and he was almost up to date on all that was happening in the development of mathematics and science both in England and on the continent. His travels had allowed him to accomplish what he had set out to do. He boldly turned down an invitation to serve as a parish priest in Strasbourg, resolving instead to concentrate on mathematics back in Basel. He knew that his younger brother Johann, always his eager pupil, was ready to work with him as they put together Jacob's latest studies in mathematics.

11
Jacob Settles into Life in Basel to Lecture and Learn

Once again, Jacob's father was impatient for his oldest son to begin his career and accept a position as a pastor in the Reformed Church, but by this time Jacob had independently decided against that move. Realistically assessing his own abilities and goals, he was satisfied that turning down the offer from Strasbourg had been a good decision.

"Jacob," his father approached him with great concern, "do I understand that you have turned down that excellent position in the church in Strasbourg without consulting me?"

"Yes, Father," Jacob replied. "As you know, I have been working diligently on mathematics for several years, and I can't stop now. I am working at the forefront of mathematics today, and I must continue."

"Now, wait a minute, young man," his father said. "I cannot accept this. I was willing to let you travel after you completed your studies, but it was always clear that afterwards you would accept a position in the Church and make it your career. I know you understood that. Otherwise I would never have allowed you to go on your travels."

"I'm sorry, Father," Jacob began carefully. "I realize that you planned for me to devote my life to preaching the gospel, but instead I have found my own calling: mathematics. Like Martin Luther, I must follow my own calling. Perhaps you didn't know that Martin Luther's father had planned for him to become a lawyer. He must have been aghast when Luther instead chose a life in the Church. But you will recall that Martin Luther boldly said, 'Here I stand. I can't do anything else. God help me.' I say the same to you."

"No!" his father barked. "I have worked for years preparing you for your distinguished career in the Church. I feel as if you had just slapped me in the face."

"No, Father" Jacob said sadly. "This is not intended as an insult to you. I am truly sorry that you can't understand my passion for mathematics. You should know that when I am exploring mathematics, I do it with a near religious fervor—this is not a mere whim. This will be my life. Please don't condemn me for it."

"No! Religious fervor is for religion!" his father declared.

"If a man feels passionately about his vocation, it is his religion, Father," Jacob said. "I am a scholar. God has chosen me to pursue an

Jacob Bernoulli.

understanding of our world at a fundamental level. I have met with all the great mathematicians in the world, I have won their respect, and now I must join them in their work. Mathematics provides the foundation for all of science, and I must play my part. I have no intention of forsaking the Protestant religion, but that is not where I will make my career."

Jacob's father sighed. "You are a foolish young man. You are throwing away a brilliant career where you would have been respected universally. I can't believe it."

"No, Father," Jacob corrected him, "I am throwing nothing away. I fervently hope I will have a brilliant career, but it will be in mathematics, God willing. I have been working diligently toward this goal for several years, and I have no intention of stopping now."

In 1682 at the age of 28, Jacob decided to publish in the *Acta Eruditorum* [*Acts of the Scholars*], a scientific journal from Leipzig, Germany, his discoveries about comets and their orbits. He had discussed his research with scientists that he met on his travels, and now was the time to publish it.

At about the same time he published another article *De gravitate aetheris*, concerning the weight in the atmosphere of the aether, the mythical substance that many scientists of the time thought explained such phenomena as gravity. He wrote that it is obvious that air has some weight since we can measure atmospheric pressure with a barometer. He noted that he agreed with Malebranche, his host in Paris, who also doubted the existence of the aether, although neither Malebranche nor Jacob had a good alternative to explain the mysteries of the universe. In his article, Jacob argued for the wisdom of geometry and physics, which he thought between them were far more likely to produce a plausible explanation of the physical world than the mysterious aether. Certainly there was nothing more than circumstantial evidence for aether's existence anyway. Jacob considered this still a work in progress and he eagerly awaited the next development.

In 1683, Jacob presented himself to the citizens of Basel as a lecturer in physics, offering lectures on the experimental mechanics of both solid and liquid bodies. His brilliant lectures, which were marked by clarity and enthusiasm, quickly became so popular that Jacob was soon earning a significant amount of money from his teaching.

"Heinz, my friend, a hearty good morning to you!" 25-year-old Peter greeted his friend at the Basel city hall one morning. "Did you see this notice?"

"I was just looking at it," Heinz said. "That is a lecture that young Jacob Bernoulli is offering on mechanics and physics. It sounds most intriguing. He asks only two *Pfennig* for the lecture tomorrow evening, and I believe I will attend."

"What a good idea," Peter responded. "How would it be if I stop by your house tomorrow at five o'clock and we go together?"

"What fun that will be!" Heinz said. "I hear that *Herr* Bernoulli has been learning about many fascinating things on his journeys."

The following evening Peter and Heinz walked together to the community hall where Jacob would be speaking. When they arrived, they saw Jacob setting up his equipment in the front of the hall, testing his apparatus carefully to be sure that everything would function perfectly.

"Good evening, gentlemen," Jacob began when the crowd quieted down. "I am pleased to see so many of you for my first lecture. I am planning to do a series of five lectures, each on a different topic of mechanics. Before I begin, I have only one request: if you cannot hear me or if what I am saying is not clear, please interrupt me immediately. I will do my best to answer any questions you may have. When I finish, I hope you will all be willing to leave the two *Pfennig* that I have requested for your tuition on the table here at the front of the hall.

"My topic this evening," Jacob continued, "is capillary action. Are any of you familiar with the term?" Jacob scanned the audience and saw only looks of curiosity. "Capillary action is something that you all have witnessed. Consider this dry cloth that is touching this puddle of liquid. Look closely and you will see that the liquid is slowly seeping into the cloth. Isn't it odd that it can move across, not just down?

"Now, please observe this narrow glass tube which I have inserted vertically in this vessel of water," Jacob continued. "Notice that the liquid is rising in the tube—it is going up, not down. It is the same phenomenon: capillary action. Now consider for a moment, please, the quill that you sometimes dip in ink so that you can write on a document. What keeps the ink in your plume so that you can write several words between dips in the ink? It is the very same phenomenon: capillary action."

"Pardon me, *Herr* Bernoulli," one of his listeners called out, "could we please see that demonstration with the tube one more time?"

"Certainly," Jacob said, removing the tube from the water and shaking out the remaining liquid. "Now I insert the tube once again, holding it steadily upright. Do you see that the liquid is once again rising?"

"Thank you," his questioner said with satisfaction.

"Now there are several variables we need to consider," Jacob continued, as he explained about the difference in the quality of a liquid—oil or water or mercury—and his audience could see that the capillary action was different in the more viscous liquids. Then he proceeded to show them the effect of a wider tube as compared to a narrower tube, before he moved on to a scientific explanation of why it worked.

Next, he proceeded with an explanation of the measurement of barometric pressure and the way capillary action allows us to measure the pressure of the air in the atmosphere on a pool of mercury. He explained that the height of the column of mercury depended on the pressure exerted by the air on the pool of mercury in which the vertical tube was standing.

As he talked, there was an occasional gasp of wonder at a new revelation, but otherwise by now the lecture hall was silent. His audience did not want to miss a single trick. At the end of the demonstration, Jacob announced that he was indebted to Robert Hooke (1635–1703) of the Royal Society of London for parts of his demonstration. "I spent some time in London talking with Mr. Hooke," Jacob explained, "and he seemed pleased to show me some of the devices he has made. I have seen his famous book *Micrographia*, a beautiful volume with amazing drawings of the microscopic world." Jacob then explained further that in fact the first functioning barometer was built several years earlier by an Italian named Torricelli in 1643.

When the lecture was over, Jacob announced that his next lecture would be the following Thursday evening at the same time in the same place. His topic then would be the process of combustion—a topic that Robert Boyle (1627–1691) in England had done some fascinating work on. He explained that he had seen Boyle's demonstrations and had been amazed. He thought his listeners would have the same reaction. Everyone gladly left the money on the table for Jacob, and several asked if they might try Jacob's experiments with the glass tubes for themselves. Jacob supervised them carefully as they saw for themselves how capillary action works. "Look at this, Heinz," Peter said to his friend. "The water really is climbing up the tube! I wouldn't have thought it would be possible!"

"God in heaven! So it is!" Heinz agreed.

As they left, Peter thanked his friend warmly for encouraging him to attend the lecture. "Heinz, you were certainly correct that young Bernoulli's lecture would be fascinating. He has a real knack for explaining difficult things, and I think it would be safe to say that he has truly seen the world. I had never worried about why a rag absorbed water if it wasn't even submerged in the water, or why the ink stays in my plume as I write. This was fascinating!"

"I believe *Herr* Bernoulli has been studying with the most important scholars in Europe," Heinz said. "As he was talking, I was reminded of times when I was a schoolboy and learned something exciting and new. What could be more fun than that?"

This was popular education for ordinary people—people who were becoming aware that there were exciting developments in the world of science. It turned out that many people were willing to pay for the privilege of hearing a knowledgeable scientist speak, particularly once they realized that they could understand what he was saying. Jacob's father may have finally admitted to himself at this time that perhaps Jacob's mathematical and scientific studies had not been so foolish after all.

In 1684 at the age of 30, Jacob married Judith Stupanus, who, like Jacob, had grown up in Basel. She was the daughter of a successful businessman in town. Jacob and Judith had two children—a son named Nicolaus (after his grandfather) and a daughter named Verena—but neither of these children chose to study mathematics or physics. Jacob's son Nicolaus became a painter like his uncle Nicolaus, and his daughter married a successful businessman.

In 1687 Jacob devised a method for dividing a scalene triangle into four equal parts geometrically with a pair of perpendicular lines. His friend Jean Christophe Fatio-de-Duillier from Geneva had sent this challenge to Jacob after learning of it from the esteemed Dutch mathematician Huygens, and Jacob was able to accomplish it through a remarkably skillful manipulation of Descartes' geometry. Jacob was pleased to publish this result and with it to win further respect from the scientific community.

Also in 1687, four years after his return to Basel, 33-year-old Jacob finally was chosen for the chair of mathematics at the university in Basel. Now he would be recognized as Professor of Mathematics. The long years of standing up to his father's pressure to make the move into his "real" career in the Church had finally paid off. However, as a professor, his salary was almost as small as his father had predicted many years earlier.

For the rest of his life, Jacob had to supplement his salary through private tutoring in addition to the fees he earned from his popular extracurricular physics lectures. As a serious professor of the latest mathematics, he was soon attracting students to Basel from throughout Europe. A succession of these students, who had heard of Jacob's reputation as a brilliant teacher, lodged with the Bernoullis, paying for their professor's hospitality as well as his tutelage. This was an additional burden for his wife Judith as well, but they were both committed to Jacob's career.

Beginning in 1690, Jacob's lectures in physics and mechanics were listed in the university catalog, with an official meeting time on Thursday afternoons at 3:00. By this time, his lectures had become so popular that their location had to be changed—there wasn't enough room for all the eager listeners in the original location. Now they had to be conducted in the dining hall of a music school nearby—a sure indication of Jacob's success.

One of the students who may have lodged at Jacob and Judith Bernoulli's house at this time was a poor but very bright young man named Paul Euler (pronounced *oiler*). Euler's brilliant son Leonhard would later become an extraordinary student of Jacob's younger brother Johann. Paul Euler was preparing for a career in the Church, but he was intrigued by what he had learned in mathematics, and he eagerly studied with Jacob Bernoulli. At the time, Paul Euler also came to know Jacob's brother Johann, since they were about the same age, may have lived for a time under the same roof, and frequently listened to the same lectures. In 1688, Paul Euler was a successful respondent to Jacob in a series of disputations on ratios and proportions. From that time on, the connection between the Eulers and the Bernoullis was always close.

In the early 1690s, Jacob Bernoulli discovered a revolutionary way to graph a point or an equation on the plane in a way that was totally different from Descartes' method. His brother Johann, who was now in his early twenties, was eager to hear all about it. At this time, they often worked together on mathematics—Johann didn't want to miss anything that his brother found interesting.

"Now, Johann," Jacob began, "take a look at this new method of graphing that I have just come up with."

"But I thought Cartesian graphing did everything we would need to do," Johann protested. "I've never had trouble with it."

"I think my new method is even better than Descartes'," Jacob said. "From the origin, I'm going to draw a ray going off to the right, technically forever, although we would never draw it that way."

"So far it sounds just like Cartesian graphing to me," Johann complained.

"Let me continue!" Jacob protested. "You may think you know everything, but I still have some advantages over you. Hear me out! What I have discovered is truly revolutionary."

"All right," Johann said meekly, "I'll listen."

"Okay, so we have this ray with one endpoint at what Descartes called the origin," Jacob said. "The new way to locate a point is by taking two measures: the angle away from that first ray measured in terms of π, and then the linear distance out the ray."

"Aha!" Johann said, suddenly seeing where his brother was going with this. "So if we want to graph a point on the Cartesian y-axis above the origin, we would describe the 90° angle as $\pi/2$, since π would be 180°—aren't we saying that 2π would be one complete rotation? Then we would measure how many units away from the origin you need to go to reach the point."

"That's right," Jacob said. "Then the ordered pair of that location in polar coordinates would be $(\pi/2, 3)$ if the point is three units above the origin. So the first entry of the ordered pair gives the angular measure, and the second gives the distance. That's all we need to know."

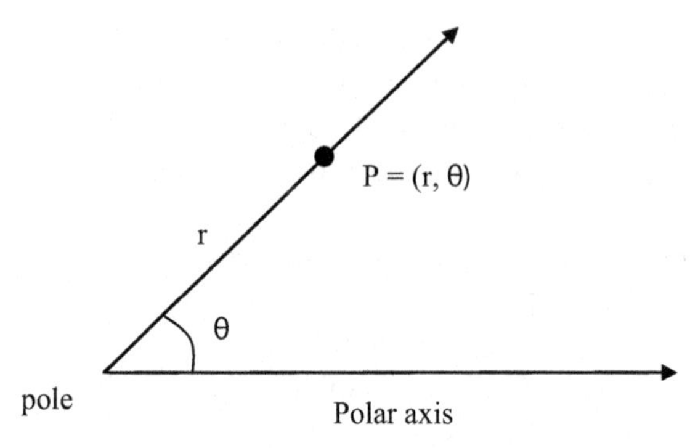

Graphing with polar coordinates.

"Yes, and that would be the same as the point (0, 3) in Cartesian coordinates," Johann said. "I think I like this. Is it original with you? Did you make it up yourself?"

"I devised it myself, and as far as I know I am the first person to do it," Jacob said. "I'm in the process of writing it up for an article in *Acta Eruditorum*."

"What advantages do you think this new method has over Descartes' method?" Johann asked.

"Well, for one thing," Jacob said, "it allows us to consider motion—not just things that are stationary. For another, we could represent a given point by indicating that it had been rotated 1/4 of the way around, or we could describe that same point as 9/4 or even 17/4. I am finding it remarkably useful."

"I like it, Jacob," Johann said. "Do you mind if I try using it in my work?"

"No, I'd actually like you to use it," Jacob said enthusiastically. "Please tell me if you discover anything more about it. You truly are my best critic, and I value your opinion."

Jacob published his method of polar coordinates in the *Acta* in 1691, and it created a major sensation among the mathematicians of Europe. In fact, Johann was later surprised to learn that Jacob was not the first person to use it. Isaac Newton had come up with the same scheme several years earlier, but, typically for Newton, he didn't publish it until 40 years later in 1736. Since Jacob published it first, he is the one who deserves credit for it.

At this time, the two Bernoulli brothers were happily working together on their mathematics. Jacob had started ahead of Johann, but Johann always grasped the new concepts so quickly that they were truly operating as equals in most ways most of the time. Although they fought brutally when they were older, that strife was still in the future at this time. They were still two congenial adventurers blazing new paths into the wilderness, making many exciting new discoveries as they worked. It wouldn't have been half as much fun without an accomplice.

In addition to his explorations into mathematics in 1691, Jacob foolishly involved himself in university politics.

"Professor Schmidt," Jacob addressed one of his colleagues one afternoon, "what do you think of our policy of allowing a professor to teach in a field that is not his area of expertise?"

"Well, Professor Bernoulli," his colleague replied, "I must admit that I have never worried about it. We have many fine scholars on our faculty."

"But think about it, Sir," Jacob persisted. "Our students come to the university to learn from scholars in their fields. A professor of law who claims to be a scholar of Greek is a fraud as I see it!"

"Now, wait a minute," Professor Schmidt corrected him. "That is the way the university has always functioned. I don't think we need to disturb the workings of the university in an attempt to be purists. In general, you must agree that our program is excellent."

"But don't we want to present the best scholarship that we can to our capable young students?" Jacob asked. "Shouldn't all of us be the best scholars we can possibly be?"

"I must say that I would not want to make an issue of it," Professor Schmidt admitted. "Wouldn't it be better for you to simply concentrate on doing the best you can in your own field?"

Jacob, ignoring the advice of his older colleague, spoke to others on the faculty and even to some officials of the university, causing considerable unpleasantness, and eventually resulting in a suspension of his position on the faculty. He was justifiably perceived as trying to stir up unrest. Whether or not Jacob was right, those in positions of power found the current arrangement thoroughly satisfactory and were unwilling to see the system that served them so well turned upside down.

Fortunately for Jacob, his father had enough prestige in the community to step in and push the authorities to reinstate Jacob. In fact, Jacob may have been disingenuous in his stance since when he had been trying to join the faculty as a professor only a few years earlier, he had proposed twenty theses to defend in a wide variety of fields (not just mathematics). "Would anyone like to hire me as a professor of moral philosophy?" Jacob had asked. In later years, future Bernoullis (his nephews and great-nephews) were also guilty of this same "offense." Although Jacob was probably right that the university policy was not ideal for academia, it was not a battle that he was going to win for many reasons, and the policy certainly helped more than one Bernoulli over the years.

12
Leibniz's Calculus vs. Newton's Fluxions

Twenty-five years earlier, in the years 1665–1666 and far from Basel, the 23-year-old scientist Isaac Newton was a refugee from Cambridge University on his family's farm in central England, far from the highly contagious disease called the plague, which had forced the closing of the university until the danger passed. Newton spent those 18 months thinking and discovering and experimenting, in what has since been called his *Anno Mirabilis* [Miraculous Year]—the months when he made more brilliant discoveries in science than perhaps any other single person has ever done in so short a time. His only restrictions were the limits imposed by his own imagination and curiosity—and these were amazingly vast and deep.

Newton had retired to the country where he was at liberty to think and experiment and pursue his discoveries wherever they might lead him, free from any cares. Looking back on that time, the older Newton said, "In those days, I was in the prime of my age for invention and minded mathematics and [natural] philosophy [meaning science] more than at any time since." Newton was so consumed with his research that he did little else during those months, often forgetting to bathe and sometimes even to eat or sleep. There is a story that his cat grew luxuriously fat from eating all the untouched food that was set out for the possessed young scientist.

It was during this time that Newton came up with the basic concepts for the part of mathematics that he called fluxions and that we now call the calculus. Newton's brilliant insight in mathematics was to see that further exploration of algebra and geometry must center on motion. He saw a curve not as a collection of points, as in Euclid's classical geometry or Descartes' analytic geometry, but rather as movement and change. His fluxions (related to the words fluent and flowing) were a dynamic study, in which he looked at the instantaneous speed of a particle and the area found under its curving path. He did that by constructing the ratio of the distance covered in the journey to the time it took to cover it. His method was to look at both those measures as they were reduced to the smallest possible increment—not zero, but very, very close to zero—what is now called the infinitesimal or the infinitely small.

Archimedes (287–212 B.C.) had approached the infinitesimal more than 1800 years earlier, as had many others since then, but there is no doubt that Newton was the first to see how to use it to solve a wide range of problems. However, he would have been the first to admit that his new explorations in mathematics were not an isolated piece of work carried out by him alone—he was one actor in a continuum of scientific discoveries. By this time, he was already well grounded in the mathematics that had been discovered over the preceding centuries, and the field was ripe for further development. He said once that he had accomplished all that he had because he was able to stand on the shoulders of giants—of Archimedes, Huygens, Descartes, Wallis, etc. The mathematical world was ready for the discovery of the calculus, and Newton was the first to put it all together. Because of Newton's amazing accomplishments in mathematics, he is today considered one of the four greatest mathematicians of all time, coming after Archimedes and before Leonhard Euler (1707–1783) and Carl Friedrich Gauss (1777–1855).

Newton's discoveries during those 18 months were not limited to mathematics alone. Stories tell of his experiments in optics and vision. Once he looked straight at the sun for as long as he could stand it in order to understand the way the eye works, although he paid for that experiment later as he was forced to spend several painful days inside a dark room while his eyes returned to normal. In fact, he was lucky that he didn't blind himself completely at that time. Another time he experimented with the effects on his vision of varying the shape of his eyeball by inserting what he called a bodkin [a large, blunt needle] as far behind his eyeball as he could in order to observe the effects on his sight of a change in the curvature of his retina. That experiment could also have had disastrous results, but once again the ingenious scientist escaped unharmed.

At this time, Newton also experimented with refracting a ray of light into the spectrum of colors from violet to magenta using a prism, although people argued at the time that this was nonsense. It was common knowledge that normal daylight, which is clearly white, couldn't possibly be composed of all those colors! While Newton knew he was right about this and his other discoveries, he felt no strong desire to convince anyone else of that. He was apparently content to explore solely for the sake of exploration. It was relatively unimportant to him what use others would make of his discoveries.

There is another famous story (which is probably not literally true) describing Newton's inspiration at watching an apple fall to the ground, leading to his discovery of the universal law of gravity. Many of his contemporaries were critical of Newton's concept of gravity, since he couldn't explain why it worked. They enjoyed mocking his concept of a certain "drawing-ness"—as they dubbed the power of the attraction that he called gravity—between the earth and the sun or between an apple and the earth. The skeptics condemned it as farfetched, but later scientists discovered that gravity actually does work the way Newton said, regardless of its cause. In later years when Newton was hailed as a hero, he modestly claimed that he had merely been like a boy who had happened to find some particularly pretty stones while playing carelessly on the seashore.

By the time Newton reached middle age, he enjoyed the passionate respect of his colleagues and the general public throughout England. He was knighted and called Sir Isaac Newton, served for a time in parliament, became an effective and energetic Master of the Mint, and was buried in a place of honor in Westminster Abbey. Although he was not of noble birth, by the time he was 30 years old he was esteemed as the most noble scientist in the English-speaking world. While he had few friends and generally worked in splendid isolation, his genius was universally recognized in his homeland.

The story of Gottfried Leibniz is totally different, although, like Newton, he certainly was a genius. Unlike Newton, he was far more than just a scientist—he was a polymath. He was a savant who worked brilliantly in many fields, from law to philosophy to mathematics, and who, unlike Newton, also enjoyed communicating with others. Newton enjoyed his own company far more than the company of others.

Leibniz's greatest regret was that he was not a nobleman by birth, one who could enjoy the privilege of making witty conversation in the courts of Europe throughout his life just because of who he was. He would have loved to devote his time to any intellectual pastime that he chose for as long as he wished, as Newton had been able to do throughout his adult life.

In spite of his remarkable accomplishments, however, Leibniz was never awarded the status of a nobleman as Newton was. When Leibniz died, he was buried in an unmarked grave, unrecognized and unsung, with only his former secretary in attendance at the interment. The contrast with Newton seems grossly unfair.

Leibniz's father had been a professor of philosophy who possessed a large library in which young Gottfried was allowed to read widely after his father's death. Having an insatiable curiosity, the boy began doing this from a very young age. Although he attended the local school beginning at the age of seven and was instructed there in Latin and Greek, he had already taught himself those languages in his desire to read all the books in his father's library. There was no stopping that child! Leibniz went on to study at the university in Leipzig, successfully completing his Bachelor's degree at the age of 16 and his master's degree in philosophy a year later.

However, when Leibniz applied for his doctorate in law, the university at Leipzig refused, perhaps because they considered him too young, but more probably because they were limited in the number of doctorates they could award in a given year. They apparently reasoned that since Leibniz was only 20 years old, he could certainly wait another year for his doctorate.

Not one to accept defeat, Leibniz promptly traveled to the nearby university at Altdorf where he submitted his brilliant dissertation and was soon granted his doctorate in law there at the age of 21. Clearly a gifted and accomplished student, he was immediately offered a professorship in law at Altdorf, but he promptly turned that down. A provincial university was too small a setting for him.

Leibniz then attached himself to a series of noblemen who appreciated and were eager to exploit his brilliance. As an expert in the law, Leibniz had much to offer, and he was pleased to make himself valuable to noble sponsors at the same time that he saw the world. In this way, he was able to take part in the world of nobility, even though he was not personally a member of that club. While he worked for one of these noblemen on a diplomatic mission in Paris, Leibniz was delighted to discover the world of mathematics beyond rudimentary reckoning. Studying seriously under the guidance of the Dutchman Christian Huygens, the most important mathematician of his time, Leibniz was enthralled, and with his genius he was able to progress rapidly.

He then traveled to London, also on a diplomatic mission, and there he met with members of the Royal Academy, in his free time demonstrating his brilliant plans for a calculating machine that he claimed would add, subtract, multiply, divide, and take square roots. It was an inspired idea, although the many delays in the actual construction of the machine caused him considerable embarrassment with his London contacts over the years. In fact, the machine never functioned as he had planned.

While Leibniz was in London on two separate trips, he heard of Newton, although he did not actually meet him. It is possible that at the time he saw at least one privately printed piece of Newton's work on fluxions, but, if he did, there is no record of that event. With his limited mathematical background at the time, he probably could not have understood Newton's writings even if he had had the opportunity to study them carefully.

When Leibniz later returned to the continent and explored mathematics in his occasional free moment, he began to see the need for the analysis that might be possible with some new mathematical tools. He then wrote to Newton asking for some information on his work, which Newton eventually sent to him, although he encrypted it so thoroughly that Leibniz was able to learn nothing from it. Newton's goal was to establish his priority without actually divulging anything. Since Leibniz couldn't decipher it, neither of them gained anything from that correspondence.

Seven years after Newton's discovery of his fluxions, Leibniz once again found himself on a diplomatic mission in Paris between 1672 and 1676. It was during those years that he discovered his calculus. Leibniz's creativity was stunning. He devised the notation that we use today: $f(x)$ and dx and $\int$, compared to Newton's fluxions, which used such symbols as x with one dot above it or x with two dots above it for the first and second derivatives. Since Leibniz's calculus is the version that we use today, mathematicians prefer Leibniz's calculus notation, although that may be simply because it is familiar to us.

Leibniz's calculus came to be called the calculus as a shortened version of the title of his 1684 article in *Acta Eruditorum,* the scientific journal that Leibniz had helped to found and where Jacob Bernoulli published his first discoveries. The Latin title of Leibniz's article was *Nova Methodus pro Maximis et Minimis, itemque Tangentibus, quae nec fractas nec irrationales quantitates moratur, et singulare pro illis* calculi *genus* ["A new method for Maxima and Minima as well as Tangents, which is neither hindered by fractional nor irrational quantities, and a remarkable type of *Calculus* for them"]. The word *calculus*, found in the word *calculi* at the end of that title, is the Latin word for *pebble*, referring to the pebbles in an abacus used for calculation. The word fluxion is seen today as no more than a quaint reference to Newton's system that did not win out. Today we study only Leibniz's calculus.

Newton did not publish anything about his fluxions until 1687, many years after his own discovery of it and three years after Leibniz's first publication of his calculus. Even then, Newton's method of fluxions was only an incidental part of his monumental work *Philosophiae Naturalis Principia Mathematica* [*Mathematical Principles of Natural Philosophy*] which is usually called simply *The Principia*. In it he used his fluxion method occasionally, although most of his proofs used only traditional geometry. The work includes no clear presentation of Newton's method of fluxions.

Unlike Newton, Leibniz eagerly published his calculus, but his first article on the calculus in *Acta Eruditorum* was a mere six pages of dense and exotic calculations with very little explanation. During his youthful travels, Jacob Bernoulli had heard something of Gottfried Leibniz, an impressive scholar in mathematics and many other fields and one of the founders of the journal *Acta Eruditorum*, which Jacob now read regularly. Since Jacob had already mastered all that Huygens and Descartes and the other great mathematicians of Europe had presented, Jacob knew that it was now time to read Leibniz.

After studying Leibniz's article carefully, Jacob Bernoulli described it as an enigma rather than an explanation. Other savants, who had also tried to read it, had simply given up. Leibniz's work used discoveries that he had made sometime before 1677, at least seven years earlier than the current article. Presumably this article was an improvement on that earlier work, but it was still unclear to even his most determined reader—Jacob Bernoulli. Although Jacob wrote to Leibniz in 1687, asking for some clarification, he received no answer, probably because Leibniz was traveling for his noble sponsor and thus was out of touch with his correspondence. He would find his mail when he returned home six months later. Thus, Jacob had no choice but to persevere on his own. In later years, Leibniz admitted that his calculus was as much the work of Jacob and Johann Bernoulli as it was his own. Although Leibniz certainly had the initial inspiration, he needed the Bernoullis to present it to the scientific world.

In 1691 Jacob published two essays on Leibniz's infinitesimal calculus, based on his teaching of the subject to his private students at the university in Basel and the work he and his brother Johann had done together. These essays were the first presentation of the infinitesimal calculus that were clear enough to allow other mathematicians to begin to comprehend the subject. The development of Leibniz's calculus from the end of the seventeenth century into its many forms was the major accomplishment of the eighteenth century, due in great part to the revolutionary work of the Bernoullis and then of Leonhard Euler, the brilliant son of Jacob's former student, Paul Euler.

Twenty years after Leibniz published his discovery, a war ensued over who should get credit for the development of the analysis that we call the calculus, the branch of mathematics that first allowed us to find the instantaneous velocity of a particle and the area contained within a curve. The priority battles continued long after the deaths of Newton and Leibniz, effectively cutting off English mathematicians from continental mathematicians. As a result, mathematics developed separately in England and on the continent for at least a century. In great part because of the work of Jacob Bernoulli, his brother Johann, the Bernoulli brothers' students, and Leonhard Euler, the continental scholars were able to carry their investigations much further and much faster than their English counterparts, with the result that the English lost any competitive edge they might have had. Because Leibniz's mathematics was the active medium for the development of later mathematics, Newton's notation and English mathematicians lost out. That is the reason that fluxions are no longer a part of standard mathematics.

The war of the two calculuses is complicated in many ways. Newton certainly developed the analysis first but didn't publish it until much later. Leibniz may have seen some suggestions of what Newton had done when he was in London, but Leibniz's method was original, and he published it several years before Newton. The Newtonians who accused Leibniz of plagiarizing were wrong. On the other side, some continental mathematicians boldly accused Newton of plagiarism, saying that Newton could have read Leibniz's calculus, which had been in print before he presented his own. Both Leibniz and Newton are now respected as having independently discovered the calculus. Neither of them was guilty of plagiarism.

13
Johann Bernoulli Grows Up

In 1683, with Jacob lecturing in Basel on topics in physics but before he was named professor, his father decided that it was time to make preparations for his younger son Johann's career. He had lost the career battle with Jacob, but now he had another chance. Johann, a remarkably bright young man, had completed the standard schooling, and his father decided that with his keen mind the ideal career for him was in business. He saw Johann not as a brooding young man like Jacob, but rather as one with a quick wit and the perfect personality for a life in business.

"Johann," his father said to him one morning, "I have arranged an apprenticeship for you with a very successful businessman I know in the town of Neuchâtel. He expects you to arrive next Monday to begin your work with him."

"What did you say I am going to do?" Johann asked in horror. "Am I supposed to become a businessman?"

"Yes," his father said. "You are a capable young man who learns new things easily, and I believe you have the perfect personality for this career."

"But I wanted to study at the university, like Jacob," Johann protested.

"No, Johann," his father said, "you and Jacob are very different people. You have a keen and penetrating mind, and so a career in business is right for you."

"You may be right that I'm smart enough for this, but what if I don't want to become a businessman?" Johann asked.

"You will accept this apprenticeship," his father announced. "I would recommend that you take a little time this weekend and review your French studies so that you will be able to communicate with your host easily when you arrive."

It soon became clear that Johann had even less interest in becoming a businessman than his older brother had had in becoming a pastor. One year later, after many protests, Johann was finally allowed to return home to Basel to study at the university there. His father had lost once again. In 1685 at the age of 18, Johann stood in a debate against his brother Jacob at the university, with the result that Johann was granted the degree of Master of Arts so that he might begin the study of medicine, his father's second choice of a career for him. In 1690 at the age of 23, Johann passed the licentiate in medicine with a thesis on fermentation, a decidedly mathematical piece of medical research.

After he published this work, Johann quietly broke off his study of medicine for a few years. Mathematics was his interest, and he pursued it eagerly. He was determined to learn whatever mathematics his brother Jacob had learned, and soon they were operating at the same level, although Jacob later described Johann scornfully as his student, who, as Jacob had predicted many years earlier, would never be able to do anything in mathematics unless Jacob chose to teach it to him. In fact, that is not the way it happened, as their relationship became complicated in several ways. Nevertheless, they were both formidable mathematicians whom the mathematical world quickly came to respect, even when they showed little respect for each other.

Jacob and Johann, who were both working hard at understanding Leibniz's writing on the calculus, were the first people to genuinely understand the details and the potential breadth of its applications. Although Leibniz had discovered it and used it in a limited way, and Leibniz's friend Ehrenfried Walther von Tschirnhaus

(1651–1708) had explored some of its applications, neither Leibniz nor von Tschirnhaus had developed a clear presentation of the material, and neither of them had been able to generalize the techniques. While Leibniz and von Tschirnhaus had consistently limited themselves to solving specific problems, the Bernoullis could see that the work was important in a much broader way.

Jacob and Johann carefully read all the works of Leibniz and von Tschirnhaus that were published in the *Acta Eruditorum* between 1682 and 1686.

"Johann, look at this paragraph from Leibniz," Jacob said one afternoon. "He is looking for a good way to find the slope of a curve at a specific point."

"But Jacob, isn't that what Barrow did?" Johann asked. "Barrow was able to find the slope of the line using Descartes' geometry."

"Yes, Barrow did that," Jacob said, "but it's possible that Leibniz is taking this much further or maybe in a different direction. Let's reserve our judgment until we see where Leibniz is going with this."

Several minutes later Johann observed, "Leibniz certainly didn't worry about making it particularly clear."

"No, and I've written to him for clarification, but so far he has not responded," Jacob said. "However, we're smart enough. We should be able to figure this out. If Leibniz could do it, so can we."

Since the Bernoulli brothers knew nothing of Newton's work (which had not been published), their only source was Leibniz. As they continued working, they found that in fact Leibniz had taken it much further and deeper than Barrow had, and the more they worked the more enthusiastic they became.

"Yes, Johann," Jacob said a few days later, "I think Leibniz has done something completely original here. I don't think Barrow could have done this."

"No," Johann said, "I think you are right. I think Leibniz has an entirely new method of analysis that we can use in some fascinating new ways. This is exciting!"

Without any help from Leibniz, the two brothers deduced how his calculus worked, and they were amazed at how powerful a tool it was. Jacob was soon teaching the calculus to his private students at Basel, and both brothers were moving ahead with their researches. They both understood that they were standing at the beginning of a brand new field of mathematics, and they were eager to move ahead with it

Johann spent most of the year 1691 in Geneva, teaching differential calculus to Jacob's friend J. C. Fatio-de-Duillier at the same time that he worked seriously at deepening his own understanding of it. Several years later, Fatio's younger brother Nicolas would play an active role in the debate between the Leibniz camp and the Newton camp as they struggled to establish who deserved credit for the first discovery of the calculus. Leibniz's primary defender turned out to be Johann Bernoulli, who fought for Leibniz's side energetically for many years. For Johann, defending Leibniz's calculus was a crusade that must not be lost, while the English, with help from Nicolas Fatio, were similarly fervent. It is unlikely that Johann's tutoring of the older Fatio played a role in this dispute.

At that age of 24 in the fall of 1691, Johann went on from Geneva to Paris, where he was able to enter into serious mathematical discussions with Jacob's friend Malebranche and his circle of friends, who were eager to learn more about Leibniz's calculus. Johann was also in contact at the time with the Dutch mathematician Huygens, who had been in the forefront of mathematics as it evolved from Descartes' analytic geometry and who had served as Leibniz's first mathematical mentor while he was in Paris. Through his correspondence with the Bernoulli brothers, Huygens eventually became convinced that Leibniz's calculus was correct and important, although he did not use it as enthusiastically as the younger, more active Bernoullis did.

Johann's most important new contact in Paris was the Marquis de l'Hôpital, a member of Malebranche's group of mathematicians. L'Hôpital was recognized at the time as the greatest mathematician in France. The marquis was eager to learn the calculus, and Johann agreed to teach him for a very large fee, but under an agreement granting the marquis sole rights to the material. Since Johann's father was still reluctant to support his rebellious younger son, Johann welcomed this arrangement, which was to continue for several years, although he later regretted signing over his rights to the presentation of the calculus.

At first, Johann instructed the marquis in person both in Paris and at his country estate outside of Paris, but later the instruction continued by mail at l'Hôpital's request. Johann kept good records of his instruction, retaining copies of the letters he wrote to l'Hôpital in the years after he had left Paris. Several years later, when l'Hôpital surprised Johann by publishing a textbook on differential calculus, *Analyse des infiniment petits* [*Analysis of the Infinitely Small*], Johann was pleased at first. L'Hôpital mentioned Johann's name on the title page, but that was the only credit that Johann got. Whose work was it? Johann was astonished that his student had had this in mind!

Many years later, when Johann protested his rights to the calculus textbook, he had proof that in fact he was the author, not l'Hôpital. However, their agreement had been to give l'Hôpital free use of the materials, and so the textbook is still officially called l'Hôpital's book, and l'Hôpital's Rule on simplifying an expression which involves a fraction that has a zero in both the numerator and the denominator retains his name as well. Nevertheless, credit for the first complete explanation of the calculus as well as l'Hôpital's Rule should belong to Johann Bernoulli.

While he was in Paris at this time, Johann also met Jacob's friend Pierre de Varignon. Although Johann also taught Varignon the calculus, he did it not as a formal tutor but rather as a friend and colleague. Ultimately they developed a warm friendship, as evidenced by regular correspondence that continued for many years. This time there were no payments for instruction and no transfer of rights to the material.

In 1693, Johann began to correspond frequently with Leibniz, exploring with him the general principles of the calculus. Johann was to keep up his correspondence with Leibniz for many years, often keeping Leibniz informed of his brother Jacob's work as well. Jacob observed more than once that he was too lazy to be a good correspondent, although he also had serious health problems that interfered with his activities for much of his life. However, Jacob came to resent that correspondence.

"Johann," Jacob complained one day, "is that a letter from Leibniz to you?"

"That's right," Johann said. "I had written to ask him about that problem we were working on a couple of weeks ago, and he has just responded."

"I ask you, Johann, do you consider Leibniz your exclusive friend? You wouldn't have known anything about him if I hadn't introduced you, and I resent being left out of your communications. Leibniz must believe that in corresponding with you he is in communication with both of us, but that begs the question."

"Oh, well, I just wanted to get his reaction to what we are doing," Johann explained.

"But you didn't include me in that communication, did you?" Jacob asked.

"Well, no, but I assumed that if you wanted to communicate with him, you would write a letter yourself," Johann said.

"I resent your attitude. From now on, I would like to see letters you send to Leibniz, and I would like to read his responses. You owe me no less than this."

"Why don't you write to him yourself?" Johann asked as he stormed out of the room. "I'm not your secretary."

At this time, Johann was also regularly submitting articles both to the *Acta Eruditorum* (the journal Leibniz had founded) and to the

Journal des Sçavants, another scientific journal, which had been published in French since 1665. As a result of his writing, Johann was increasingly being regarded as a serious mathematician and not just as his brother's student.

In 1694 at the age of 27, Johann finally completed his doctoral thesis in medicine on the functioning of muscles in the human body—a decidedly mathematical exploration of a medical topic. Although it was a doctorate in medicine, since that was what his father demanded, there was no doubt that Johann was a mathematician. Ten days after completing his doctorate, he married Dorothea Falkner, the daughter of one of the city deputies. Johann was ready to begin his career, preferably in a comfortable position at the university in Basel. Unfortunately for Johann, however, the chair in mathematics was already inconveniently filled by his brother Jacob.

Instead, Johann reluctantly accepted a position as engineer for the city of Basel. The job was neither interesting to Johann nor well paid. As a result, Johann was desperate to find something else.

One evening about this time, Johann's wife Dorothea said, "Johann, I'm so proud of you. You are now a recognized scholar: Dr. Bernoulli!"

"Yes, Dorothea, it took me awhile to get to this point," Johann admitted, "and I have to admit that it's a little unclear where I should go from here. Certainly I have no desire to continue in this position of city engineer. If only Jacob were not sitting in the only chair in mathematics in Basel!"

"Well, he is older than you," Dorothea observed, "by 13 years."

"Actually, it's more like 12 1/2 years," Johann said. "He was born at the end of December."

"At this point, I don't see that it makes much difference," Dorothea said. "You are both grown men and impressive scholars, but Jacob got a head start on you. So what are you thinking of doing?"

"Huygens is exploring finding me a chair in Holland at one of the universities there," Johann said. "What would you think about moving to Holland?"

"I've never thought about it," Dorothea said. "I've always lived in Basel, and, of course, I would prefer to remain at home. I had always assumed we would live in Basel and bring up our children here. I love Basel."

"But since we don't have a proper income, Basel is difficult for us," Johann said. "I'm sure we would be able to find people in Holland who speak German or at least French, and I really don't think we have much choice. For me, a career in Basel is closed unless I want to follow one of my father's plans and enter the world of business or become a medical doctor. Obviously, I don't want to do either of those things."

"Then let's see what Huygens is able to find," Dorothea suggested. "I guess I'm prepared to do whatever you think is best."

In 1695, with help from Huygens, Johann was called to be professor of mathematics at Groningen, a major university in Holland. Twenty-eight years old and with the esteemed title of professor, Johann was officially now the equal of his brother Jacob. In that secure position he could pursue his career as a mathematician. He traveled with his wife Dorothea and their 7-month old son Nicolaus to Groningen, where Johann taught mathematics and physics successfully for ten years. It certainly would not have been Dorothea's first choice, but she was pleased that Johann's career was beginning well, and she found that Groningen was a beautiful city. As a tribute to Johann, the city of Groningen now has a square that is called Bernoulli Square.

Like his older brother Jacob, Johann was also a brilliant teacher. In his writing as well as his teaching, Johann demonstrated that he not only understood the calculus in all its details—he could also explain it to those who were not already conversant in the field. He was passionate about the calculus, and he was able to inspire in his students a similar passion for the field. He is probably responsible more than anyone else for the triumph of Leibnizian differential and integral calculus over Newton's method of fluxions, which in fact accomplished the same thing.

In later years, when he described the battle over who deserved credit for the discovery of the calculus—Leibniz or Newton—Johann referred to his boyhood study of Livy's *History of Rome*. In Book II, Livy described the scene where Horatio Cocles bravely chose to defend the bridge over the River Tiber against the approaching Etruscan army. Although all but two of his comrades had fled, Horatio stood his ground and defended the bridge against the invading forces. Johann saw himself as Horatio, bravely defending Leibnizian calculus from the arrogant, misguided English. The comparison was not lost on his readers—any educated person would have known the story of Horatio and would have understood Johann's point.

14
Two Curves Studied by the Bernoullis: The Isochrone and the Catenary

In 1659, several years before either Leibniz or Newton had discovered the calculus, Huygens had been able to establish that the isochrone (from the Greek *iso* meaning same and *chrone* meaning time) was the curve along which an object under the influence of gravity would reach its lowest point in the same amount of time from any point on the curve. The curve was in fact an already familiar curve known to mathematicians as the cycloid, a curve that is formed by the path of a marked point on a wheel, as the wheel rotates along a level path. Huygens had used the isochrone in his invention of an accurate pendulum clock, using that curve as the path of his pendulum bob.

In 1690 at the age of 36, Jacob Bernoulli was so bold as to use Leibniz's calculus—what he called for the first time "the integral calculus"—to derive the same equation in an essay in the *Acta Eruditorum*, validating Huygens' result and demonstrating the usefulness of the calculus. Leibniz read and approved of Jacob's work and adopted Jacob's name *integral* for that part of the calculus. The mathematics community acknowledged Jacob's demonstration of this effective use of the calculus to solve an existing problem in mathematics as a significant accomplishment both for the calculus and for Jacob himself.

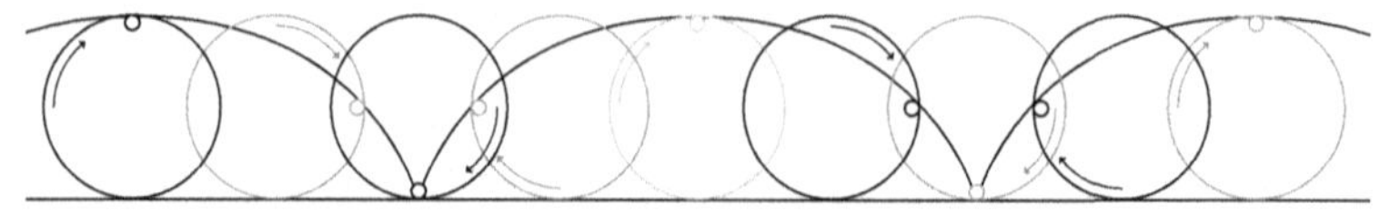

The cycloid: The path of one point on the wheel as it turns.

Having succeeded with the isochrone, Professor Jacob Bernoulli proposed a new problem in *Acta Eruditorum*, asking his readers to find the equation of the catenary curve—the curve traced by a flexible chain that is suspended from both ends and allowed to simply hang between those points. Today we see the catenary in the curve of the giant cables that support a suspension bridge. Galileo (1564–1642), who had studied that curve, incorrectly guessed that it was probably a parabola. Galileo was unable to calculate the equation of that curve because it can be done only with the use of the calculus, which had not yet been discovered in Galileo's time. When Jacob proposed the problem, he also had not yet found the equation of the catenary curve, but he still assumed, like Galileo, that it must be some kind of parabola.

In June of 1691 Leibniz, Huygens, and Jacob's then 24-year-old brother Johann (who was still living at home in Basel at the time) discovered the equation—which was not a parabola—and published it in the *Acta Eruditorum*. Jacob was mortified. His little brother, who had not yet completed his doctoral thesis at the time and who was eager to establish his own mathematical reputation, had beaten him! However, any damage Johann's discovery might cause to his brother's ego was of no concern to Johann. From this time on, Jacob and Johann's warm relationship deteriorated rapidly.

In 1718—27 years later and 13 years after Jacob's death—Johann, who was then 51 years old, was still crowing about his phenomenal success. In a letter to his friend Pierre Rémond de Monmort (1678–1719), a French mathematician with whom Johann exchanged many letters over the years, Johann described the scene in the Bernoulli family the morning after Johann discovered the equation of the curve in 1691. Johann boasted in his letter:

> Consider this, *Monsieur*! I am going to astonish you. I am telling you that my brother, try though he might, could not discover the equation of the curve of the catenary. Why should I be modest? I can tell you, *Monsieur* de Monmort, that I am the Bernoulli who discovered what that curve is. I proved that it is not a parabola, and, I assure you that Leibniz did not give me any hints. The discovery belongs to me, and I will prove it to you. You say that since my brother posed the problem, then it must be his property, but I say no. He may have posed the problem, but he couldn't solve it! Isn't that pitiful? I have to admit that at first neither of us could solve it, and we suffered. It was incredibly difficult. After all, even the genius Galileo couldn't do it.
>
> However, when *Monsieur* Leibniz announced in the *Acta Eruditorum* that he had solved it without divulging what the solution was, then the challenge was even greater. I must admit that I was awake one whole night working on this—remember my brother had been working on it for months and months without success. I have to tell you that I was suddenly fortunate—the solution came to me in a flash at the end of my long night of searching! When my brother arose the next morning I was able to present my solution to him. Poor soul, he was still miserable in his ignorance!
>
> "Stop! Stop! Jacob don't frustrate yourself any longer!" I said to him. "Don't torment yourself anymore, Jacob! It isn't a parabola, so you will never be able to make it fit the equation of the parabola. I have the solution, which I am delighted to share with you. Look at this!"
>
> Please believe me, *Monsieur*! My brother didn't have a clue what the curve was! If he had known, he would certainly have announced it to everyone. He would certainly not have allowed me to publish my result before him if he had had a choice.

> Later in my brother's correspondence with Leibniz, he indicated to Leibniz that we had solved the curve. Ha! It wasn't we—it was I! I am the Bernoulli who was awake that entire night and finally succeeded. At first Leibniz didn't know which of us had done it, but my brother and I certainly knew. Later Leibniz learned that the solution was mine, not my brother's. *Monsieur* de l'Hôpital has seen the evidence, and he agrees.

The catenary is a unique curve. Its equation involves hyperbolic geometry, a part of trigonometry that originates from the curve of the hyperbola. When the parabola is centered at the origin in Cartesian coordinates, it can be easily represented by the equation $y^2 = 4ax$ where a is the focus and x and y are the variables. The hyperbola is a basic part of algebra, while hyperbolic geometry cannot be approached without the calculus.

The catenary curve is steep at the two ends since at those points the total weight of the chain is heaviest. Toward the middle, the curve becomes less and less steep because the total weight there is decreasing where it approaches its lowest point at the curve's center of gravity. When the Bernoullis and Leibniz struggled with the curve, they were dealing with it mainly as a curiosity, but since then it has been critical to physics and engineering. The Gateway Arch in St. Louis, Missouri, is stable because it is an inverted catenary that is 630 feet high and 630 feet wide at its base. Johann used the integral calculus to reach his solution—without the calculus, the solution would have been impossible—so once again the calculus was proving its usefulness.

Eventually, Jacob too found the equation of the catenary, and he was able to present a more complete solution. Although Jacob's discovery of the catenary curve was both original and brilliant, Johann, who was eager to be seen as a scholar separate from his brother, refused to acknowledge its validity until 1715, long after Jacob's death. He would not give Jacob that satisfaction. In 1701 Johann presented his solution to the Paris Academy through his friend Varignon, but as it was presented, the solution was not complete. Jacob did not hesitate to use it as an occasion to ridicule his brother in print.

Relations between Jacob and Johann never improved after this time. They were both belligerent competitors, each determined to best the other at every opportunity. It was pointless, of course, and wasted much of the Bernoulli mathematical genius.

15
More Mathematical Challenges from the Bernoullis

In 1696 Johann, who was then a 29-year-old professor of mathematics, had settled into life in Groningen. His career was going well, he was far from his principal rival—his older brother—and he was happy.

"Well, Dorothea," he said to his wife as he returned home one evening, "I have to admit that my teaching is going very well. I am confident in the subject matter, and my students seem to be ready to learn. It is more fun than I expected it to be."

"Yes, at first you seemed a little pessimistic about the move to Holland," Dorothea said.

"Well, you were too, if I'm not mistaken," Johann said. "But isn't it fun to watch little Nicolaus as he grows?"

"He seems to be a very bright little boy," Dorothea said.

"Yes," Johann agreed. "I can hardly wait to see how he develops."

"Do you suppose he will want to study mathematics?" Dorothea asked.

"Oh, I don't know," Johann said. "Not everyone should be a mathematician, you know."

"No, not everyone," Dorothea said, "but wouldn't it be fun if he chose to do it?"

"I'm not so sure about that," Johann said.

At that time, the esteemed professor of mathematics at Groningen proposed a problem from his university office to be printed in *Acta Eruditorum*. Johann called his problem the *brachistochrone* from the Greek *brachistos* meaning shortest and *chrone* meaning time. The problem called for the discovery of the equation of the curve of quickest descent under the influence of gravity between two given points, one higher than the other and not on the same vertical line. It would be the path that a lazy hawk, wishing to coast as quickly as possible from one point at the top of a tree to a lower point on a nearby tree would fly on. Although Johann gave his readers until the end of 1696 as the deadline for entries, by the time the deadline arrived, only one correct solution—from Leibniz—had been submitted besides Johann's own.

"*Herr* Bernoulli," Leibniz replied to Johann's letter in the first few days of 1697, "I am certain that there should be more solutions to your challenge. Would you consider sending it out again, this time in the form of a pamphlet directed to the mathematicians who would have a chance of solving it? You might then extend your deadline, perhaps until Easter of this year?"

"Yes, I could do that," Johann agreed in his next letter. Then he added, "Whom do you think I should send it to?"

"Well," Leibniz responded by the next post, "certainly to your brother. And what about the Marquis de l'Hôpital? You have been working with him for several years. Do you believe his calculus is ready for that challenge?"

In his response, Johann expressed some doubts about l'Hôpital. "The marquis has learned a great deal, but I'm not sure he is ready for this challenge. However, I might as well send it to him too anyway. He would probably be offended if he knew we had omitted him from our list."

"How about Newton?" Leibniz wrote.

"Newton?" Johann wrote. "Do you think he actually has a calculus that he could use to solve my problem? I've begun to wonder if his method even approaches the usefulness of your calculus."

"We would be very foolish to underestimate Sir Isaac Newton," Leibniz warned. "The English claim that he has come up with a technique very similar to ours, and I have no reason to doubt them. Why don't you send it to him and see what happens?"

On advice from Leibniz, Johann extended the deadline until Easter of 1697. Johann sent personal copies of the challenge to each of the three other men, addressing his pamphlet to "the most brilliant mathematicians in the world." He explained that the curve was well known to geometers, and he stated clearly that, although one might wish that the correct solution might be a straight line, it was not.

The group of brilliant mathematicians was indeed a select group, and four of them were adequate to the challenge on their own. Leibniz had already solved it with little difficulty the day he received it. Jacob soon solved it as well, but with a totally different approach than Johann had used, and certainly without any help from his brother.

Newton, who had not been an active mathematician for many years, found the pamphlet in the mail when he arrived home from a long day at the London mint. He stayed up until 4 o'clock the following morning, solving it successfully using his own method of fluxions. It was a good puzzle, and Newton was not going to sleep until he had conquered it. He felt no need to announce that he had lost most of a night's sleep in solving it—his niece, who served as his housekeeper, later provided that information—although in his place Johann might have been tempted to mention that. When Newton submitted his solution, he did so anonymously. All the other entries were signed.

Although l'Hôpital, the other man on the list, wanted very much to solve the problem, he was unable to accomplish that by himself—he asked for and received generous help from Johann, whom he was still paying handsomely for his tutoring services. With Johann's guidance, l'Hôpital finally succeeded, hoping to keep his name on the list of the most brilliant mathematicians in the world.

In both *Acta Eruditorum* and the pamphlet, Johann had described the problem as a new one that he was inviting mathematicians to solve. He noted that the reward was neither gold nor silver. Instead, any successful solvers would have the supreme satisfaction and the profound respect that came to those accomplishing a great intellectual feat—a prize far more valuable than a mere financial reward.

When Johann opened the entries on Easter day of that year, he looked at Leibniz's, l'Hôpital's, and his brother's before opening the envelope from England. Although it was anonymous, Newton was the only English mathematician to whom Johann had sent the problem and he was the only person in England who Johann thought had a chance of solving the problem. Despite his prejudices, when Johann studied Newton's anonymous entry, he recognized the correct solution at once and observed, "I can tell the lion by his claw"—a comment indicative of Johann's opinion of Newton. Although Newton's notation was different from that of the continental mathematicians, it was certainly correct. In May of 1697, Johann published all five solutions (including his own) in the *Acta Eruditorum*.

Not surprisingly, the brachistochrone provided another forum for the strife between Johann and Jacob. Johann's solution involved restating the mechanical problem as an optical one—one that he could solve using Fermat's (Pierre de Fermat 1601–1665) principle of least time. Through that insight, he discovered that like the isochrone (see Chapter 14), the equation of the cycloid was the solution. Johann's was a brilliant solution to the brachistochrone, showing remarkable perception, but it offered nothing for mathematics in general. It simply solved the specific problem at hand.

In contrast, his brother Jacob constructed a much more involved argument, considering the big picture rather than the individual problem, and coming up with what turned out to be a new field of calculus—the calculation of variations—in the process. Jacob's solution was similar to Leibniz's.

Over time, historians of mathematics have concluded that both Johann and Jacob were brilliant but radically different mathematicians. Johann's wit was sharp and quick—his agile mind allowed him to see through a problem quickly and arrive at brilliant conclusions. Jacob, by contrast, operated more slowly and often came up with deeper and more general solutions.

In 1697, Jacob, who was then 43 years old, proposed another problem which he called the isoperimetric problem (from the Greek *iso* meaning same and *perimetric* meaning distance around), which asks for the determination of the curve of a given length between two points that will enclose the maximum area.

This classic question of calculus has roots in Greek and Roman mythology. In the *Aeneid*, Virgil tells the story of Princess Dido, who announced that she wished to buy land to build a city for her people on the northern coast of Africa. In reply, the wily King Jambas told her that he would sell her as much land as she could enclose in the hide of a bull, thinking that she was sure to be disappointed and would be forced to give up her plans. Dido, who turned out to be wilier than the king, had the skin cut up into narrow ribbons which were then sewn together end to end. She was able to expand the area even further by attaching the two ends of her long ribbon to two points on the seashore some distance apart so that her perimeter was even bigger, giving her the area of a half circle for her city. She cleverly solved the isoperimetric problem and was able to build her now famous city of Carthage.

When Jacob proposed the problem, it was already well known that the circle gives the maximum area if one doesn't have the advantage of Dido's stretch along the seashore. In primitive cultures around the world, circular houses have always been the favorite plan because they make the most efficient use of building materials. The problem was to prove it mathematically. The shape was easy; the proof was remarkably difficult.

Both Johann and Jacob published their solutions to Jacob's isoperimetric problem in 1701. Both men used the calculus to solve it with completely different approaches, but their end results were much the same. This time viewers from outside the family might have supposed that both Bernoullis won, but, since neither brother came out ahead, both were disappointed. Neither of them felt that he had won the contest.

The letters that Jacob and Johann exchanged at this time are full of arguments about methods of approach to several problems, pursuing the open warfare that the two brothers would engage in for the rest of their lives. This was not so much a struggle to discover new mathematics cooperatively as it was a contest to demonstrate who was more clever and more important. Over time, it appears that Johann was the one who pushed this strife more often, although there was certainly fault on both sides.

"Johann," Dorothea greeted her husband one evening as he returned home from the university at Groningen, "you have a letter here from your brother."

"Blast him anyway!" Johann exploded.

"But Johann," Dorothea remonstrated, "he is your brother. Why do you constantly fight him?"

"Because he is a lout!" Johann said. "He is a good mathematician, I admit that, but he has tried over and over to belittle my work. Why does he pursue me like that?"

"Well, perhaps you should read the letter before you get too worked up about it," Dorothea suggested. Johann tore it open and stood reading it, getting angrier and angrier.

"Yes, he's trying to minimize another one of my discoveries! I tell you, Dorothea, I hate him!" Johann said. "He likes to explain that he is the one who taught me the calculus, and therefore I can't do anything original myself. Yes, he taught me the first parts of calculus, but as time went on we worked together, each of us helping the other. He couldn't have done all that he has done without my help."

"Then why can't you work together like that again?" Dorothea asked.

"Because my brother is a rat!" Johann said, and strode out of the room.

16
Jacob Bernoulli's Mathematics

Jacob Bernoulli spent much of his life refining and expanding the calculus, which presents many occasions to contemplate the infinite. As an innovative scientist, Jacob boldly grappled with it at many levels. His poem on the paradoxes of infinity (written in Latin and translated by Martin Mattmüller, Basel, 2009) is a fine example:

Ut non-finitam seriem finita coërcet
Summula, et in nullo limite limes adest:
Sic modico immensi vestigia Numinis haerent
Corpore, et angusto limite limes abest.
Cernere in immenso parvum, dic, quanta voluptas!
In parvo immensum cernere, quanta, Deum!

Just as a finite sum confines an infinite series
And in what has no bounds there's still a bound,
So traces of divine immensity adhere to bodies
Of lowly kind, whose narrow bounds yet have no bound.
What a delight to spot the small in vast expanses,
To spot in smallness, what a joy, the immense God!

Although the significance of the calculus often eludes those who use it—the calculus can often feel more like a set of mechanical algorithms than a mind-boggling creation—Jacob didn't forget to stand back and admire the amazing mathematical machine that he was developing. He was an artist with a grand view of his vast subject.

Jacob Bernoulli's other great work—in fact his only book—is his *Ars Conjectandi* [*The Art of Conjecturing*], which was not published until seven years after his death. Jacob worked on it for 25 years off and on between 1680 and his death in 1705. He commented more than once that he suffered from both laziness and illnesses, and that both interfered with finishing his great work. In fact, the book represents a tremendous amount of original, creative work. The final product, *Ars Conjectandi,* which was eventually published in 1713, is the first complete study of the science of probability.

Jacob's first published work on that subject was a challenge that he proposed in the *Journal des Sçavants* in 1685 when he was 31 years old, two years before he was named professor of mathematics at Basel. It was written in French and directed to both intellectuals and recreational players of games. His question concerned a fictional game in which players A and B take turns throwing one die or number cube. He outlined two possible sets of rules, in both of which the winner is defined as the first player to throw a one on a standard die. With the first set of rules, in round one player A throws the die once, and then B throws it once. In the second round, player A throws twice and B throws twice. In the third round, A throws three times and B throws three times, etc. The other set of rules begins with A throwing once, then B throws twice, then A throws three times, then B throws four times, etc., continuing in this way until one of the players throws a one. The question is what is the ratio of player A's likelihood of winning to player B's likelihood of winning? Another version of this problem, later called the St. Petersburg Paradox (see Chapter 27), arose several years later when Johann's son Daniel and his cousin Nicolaus, son of Johann and Jacob's artist brother Nicolaus, were pursuing mathematics.

Unlike his article in *Journal des Sçavants*, Jacob's book *Ars Conjectandi* was not intended just for recreational mathematicians. It was published in Latin, the language of serious scientists. This time he was addressing his colleagues and students at the university of Basel as well as scientists throughout Europe—the people who read

Acta Eruditorum and who wanted to keep up on the latest developments in science. Jacob may have hoped that the book would appeal to aristocrats, too, with its obvious application to games of chance which the leisure classes had time to enjoy, but they would have to make the effort to read it in Latin.

Whereas Gerolamo Cardano (1501–1576), (the mathematician who had first attempted to teach the blind to read) wrote on probability, his work wasn't published until after his death in 1576, and it was largely ignored in the development of the theory of probability. The next serious study of probability appeared a century later in the 1650s in the correspondence of Blaise Pascal (1623–1662) and Pierre de Fermat (1601–1665), although neither of them published their results.

In 1657, Christiaan Huygens (1629–1695), the Dutch inventor of the first accurate pendulum clock and Leibniz's mentor in Paris, published his findings in his book *De Ratiociniis in Ludo Aleae* [*Concerning the Calculation of Games of Chance*]. In it, Huygens gave a method for calculating how many times a pair of dice should be thrown in order to make the probability of a given outcome worth the risk to the individual player of betting on the game. Huygens assumed that a player might say to himself: "As a rational person, I would like to know my chances of winning before I commit myself to playing." Jacob used Huygens' work as he began work on his own study.

In his *Ars Conjectandi*, Jacob presented the study of probability as an attempt at quantifying the likelihood of an event so that one could take risks intelligently. It was general knowledge in the 1680s, for example, that an insurance policy would be a poor investment unless one knew what the chances of a given event were, although at this time it was unclear how one could figure those chances.

Jacob realized that before the event, short of fixing the game there is generally no way of absolutely guaranteeing what an outcome will be. As a result, he began his study by exploring similar situations in which he could study the results *a posteriori* [after the fact] of a known event, in the hope that he might be able to predict *a priori* [before the fact] what was likely to happen in a similar situation in the future with what he called "moral certainty"—a benchmark that he would set. For this he looked back to Aristotle, whom he had read seriously in his undergraduate studies in philosophy. Aristotle had recognized that since absolute certainty is usually unattainable, an intelligent person should set a standard of certainty beyond a reasonable doubt.

A probability of one, or 100%, before an event happens is usually impossible. If the probability that it will rain is 0.5, that means that it is equally likely to rain or not rain. If the probability that it will rain is 0.9, it is much more likely to rain than not to rain, whereas if the probability of rain is 0.15, rain is unlikely. Jacob chose his standard for "moral certainty" of an event as a probability of at least 0.999. If an event's probability was 0.999—i.e., that it would happen 999 times out of 1,000—he said he could safely predict that it would happen. That was the closest he expected to come to a guarantee. After the publication of his book, his standard of "moral certainty" was soon adopted by mathematicians throughout the world.

The first section of Jacob's work is basically a summary of Huygens' book, with Jacob's own commentary bringing it up to his time. Section two summarizes the work that had appeared more recently on such topics as permutations (the number of arrangements that are possible for a certain number of events) and combinations (the number of possible sets of a certain number of events if two events occurring in different orders were considered equivalent). Section three explores the uses of probability in games of chance, and section four explores how probability can be applied to practical matters (such as the calculations of insurance premiums), moral questions (such as deciding when it is safe to conclude that a person who has been missing for several years is dead), and civic issues (for example, the construction of laws within a modern society).

When Jacob wrote about games of chance, following the example of Huygens and Pascal, he assumed that both players had an equal chance of winning—mathematicians considered those the only fair games. If two players were mismatched, it was the duty of the player with the greater probability of winning to give himself a handicap so that the chances would once again be even.

Jacob decided to study mortality statistics (the age at which specific subjects died) of a given population for which he could get the statistics, and he would use those data to calculate the mortality rate of a similar population in the future. As a result he could use that calculation to predict the likelihood of a similar individual dying at a certain age with moral certainty.

If it is likely that a person will die by the age of 40, then an insurance policy that is written for a person who is already 39 years old would need to be very expensive, while a policy for a person at age 18 would be less expensive, since he is expected to survive much longer. By the time the 18-year-old reaches the age of 40, if he has continued with his policy and paid his premiums for many years, he has already paid for his own generous payoff. It is safe to assume that the insurance company would prefer not to pay out any more money than necessary, but unless an intelligent policyholder has some chance of collecting, he would not be interested in buying a policy at all. Jacob recognized that probability does not guarantee that a given person will live to a certain age, but for the whole population it is reasonably accurate.

Jacob knew that the more statistics he could study, the more reliable his predictions would be. He decided that he could estimate the probability of a given event to any degree of accuracy he wished, using what he called the Law of Large Numbers, which appears in the last part of his *Ars Conjectandi*. While he recognized that these predictions provided only general statistics, he argued that he could make valid predictions about the population in general, and that would allow an insurance company to charge a reasonable fee for a policy.

Jacob's limited correspondence with Leibniz shows a search that Jacob pursued for many years as he struggled to complete section four of his book. In order to construct his arguments, he needed hard data, and such data were difficult to find. He repeatedly begged Leibniz to send him a copy of Johann De Witt's work, which spelled out mortality statistics from a study in Holland in 1671. He knew that Leibniz had read it, and he was convinced that Leibniz still had a copy of it. Jacob wanted it, believing that those statistics would allow him to conclude his Law of Large Numbers.

Jacob Bernoulli's gravestone in the cloister of the Basel Münster: Translated from the Latin: Jacob Bernoulli, incomparable mathematician.

Although eventually Leibniz responded to the letter, Leibniz did not send the work, saying that he no longer had it. At the time, Leibniz was still traveling extensively, researching the lineage of the House of Brunswick, his sponsor's family, as he tried to produce the history of that family. He probably didn't have the data with him as he traveled, and they may or may not have been still in his possession at home in Hannover. Without Leibniz's data, Jacob's proof was incomplete and he had no hope of finding such statistics anywhere else. In pain and fatigue, he set the manuscript aside one last time before he died.

At his death at the age of 51 in 1705, Jacob's book *Ars Conjectandi* was still not complete, but he left clear directions as to what should be done with it. Above all, he directed that it should not be put in the hands of his rival—his brother Johann—even though it might have seemed reasonable to an objective observer that the mathematical brother of a great mathematician was the logical person to see the book through to publication. Instead, Jacob directed that neither Johann nor any son that Johann might produce should even catch a glimpse of his work before it was published.

He requested that his own artistic son Nicolaus (some scholars have suggested that he designated his brother Nicolaus' son Nicolaus instead), should take the manuscript with him to Paris (where he planned to study painting) and show it to Jacob's friend Varignon who could decide if and how it should be published. Apparently Nicolaus did not do as his father asked, although he did make contact with Varignon. Jacob's wife and son held onto the manuscript and eventually turned it over to Thurneysen Brothers, publishers in Basel, and it was finally published to great acclaim in 1713, eight years after Jacob's death.

17
Johann Bernoulli Returns to Basel with His Family

Between 1695 and 1705, Johann Bernoulli and his wife Dorothea thrived in Groningen, Holland, producing several children and firmly establishing Johann's scholarly career. In 1699 both brothers Jacob and Johann were elected to the Paris Academy, and in 1701 Johann was elected to the Berlin Academy, to which Jacob had been elected several years earlier. It was clear to the academic world that the Bernoulli brothers, like Leibniz and Newton in the generation before them, were the most important mathematicians of their generation.

In 1700, Johann and Dorothea's second son Daniel, who was to become a mathematician and physicist as renowned as his father and his uncle Jacob, was born in Groningen. When Daniel Falkner, Dorothea's ailing father, realized that his grandchildren, including his namesake Daniel, were rapidly growing up far from where he could see them, he began to pressure Johann to move back to Basel. The entire Bernoulli family over several generations continued to feel a remarkably persistent tie to Basel—a tie that Johann must have felt as well—and in 1705 Johann and his family finally acceded to his father-in-law's wishes and moved back to Basel. Johann had recently been offered positions at the universities in both Utrecht and Leiden, two of the finest Dutch universities, where Johann might have preferred to relocate if the choice had been only up to him, but he had to turn those offers down.

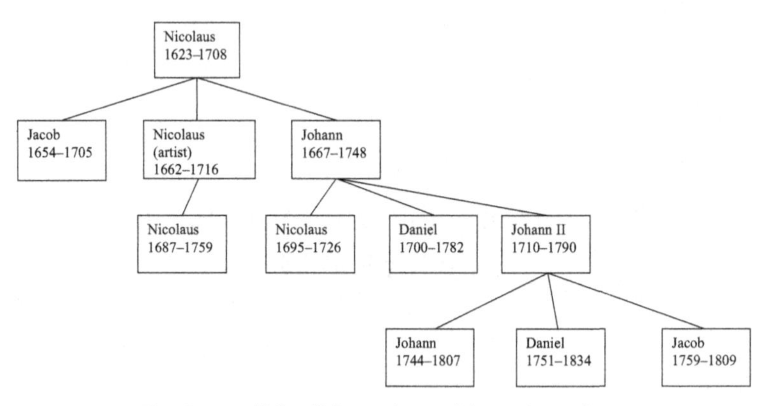

The Bernoulli family's mathematicians 1600–1850.

Unfortunately, there was no position in mathematics available for Johann at the university in Basel when he moved his family back home, although he must have been aware that his brother Jacob's health was rapidly deteriorating. As it happened, Jacob died almost as soon as Johann and his family arrived back in Basel, and Johann quickly applied for and was offered his brother's chair. Given their stormy relationship over the previous five years, this might not have pleased Jacob. Nevertheless, the university at Basel considered itself lucky to claim once again Europe's greatest mathematician as its own. Basel was not at the center of the intellectual world, but thanks to the Bernoullis it continued to earn universal respect for its mathematics.

Johann and Jacob's father was still alive in 1705 when Johann, who was then 38 years old, returned to Basel, and it seems reasonable to assume that by this time he had accepted his two mathematical sons' careers. His other sons were doing well also. Nicolaus, who was between Jacob and Johann in age, was respected as a painter, and his youngest son Hieronymus was carrying on the family business in spices.

Certainly it appeared that both Jacob and Johann had achieved the ultimate in mathematical prestige, and through a combination of their salaries as professors, their private tutoring, and renting rooms to foreign students who needed a place to stay while studying at the university, both mathematicians prospered. Contrary to their father's worries, they were not a strain on the family resources. Although Johann suffered occasionally from gout like his brother Jacob (who had been only 51 years old when he died), Johann was much healthier and lived into his eighties, extending for many years the Bernoulli monopoly of the chair of mathematics at Basel—a monopoly that would be extended yet again by his sons and grandsons.

Like his older brother, Johann Bernoulli was renowned as a brilliant lecturer in mathematics. Although he considered basic instruction in algebra little more than an annoyance and he avoided such teaching assignments whenever he could, students had high respect for the clarity of his lectures at all levels. While he was by no means rich, Johann himself was known to quietly pay the tuition for a student whom he considered worthy but for whom the cost of tuition was too high.

Johann Bernoulli and his wife Dorothea had five sons, of whom three, following the family tradition, became mathematicians. The youngest two had successful careers in business. Johann and Dorothea's four daughters, two of whom died in infancy, were expected to marry advantageously within Europe's merchant class if they were so lucky as to survive childhood. The first daughter Anna Catharina, was born in 1697 and died only a few months later, distressing her father greatly. Their second daughter, also named Anna Catharina, was born less than a year after her sister. She had a happy childhood, often playing and working with her brothers Nicolaus and Daniel, and was to grow up and marry well. Anna, as her family called her, would survive her first husband, marry again, and end up surviving her second husband as well. A younger daughter named Dorothea also prospered and remarried after the death of her first husband, this time to a pastor who later became professor of Hebrew studies at the university in Basel.

Nicolaus, Johann and Dorothea's first son and apparently Johann's favorite, was born in Basel before the family's move to Groningen, Holland. Quite naturally, Johann instructed his oldest son in mathematics from an early age since the boy seemed fascinated by it and learned it easily. At the age of eight, Nicolaus already spoke Dutch, German, French, and Latin fluently. He was clearly a very bright child, and his father Johann enigmatically chose a career in law for him.

"Father," young Nicolaus asked his father one day during the period when the family was still living in Holland, "what is the best language to speak? Which language is most important?"

"Oh, those are not easy questions," Johann said. "It really depends on what you intend to do with the language. Living in Holland, Dutch is certainly important to us."

"Obviously, Father," Nicolaus agreed, "but if we are not just talking about day-to-day use of language, then which language is most important?"

"Once again, there is no easy answer," Johann said. "As a scientist, I must speak and write Latin every day."

"Okay," Nicolaus said, "I can do that."

"Yes, you can," his father agreed, "although you will need to perfect it further as you grow older. However, Latin is not the end of my answer. In the world of modern Europe, I think most people would agree that French is the language of choice for people who are well educated. Anyone who does not speak French well will certainly be viewed as an ignoramus."

"But I can speak French," Nicolaus said, "so I'm no ignoramus."

"No, but you still need more practice with it," Johann said. "You have never actually lived in the French language, and we will have to arrange for you to do that before you are ready to go out into the world. You never really know a language until you have lived in it."

"Okay," Nicolaus said, "but what about German? We speak German at home, but is that just because you and Mother grew up speaking German? When I do arithmetic, I always do it in Dutch. I think Dutch may be the most useful language for me. You do arithmetic in Dutch, don't you, Father?"

"Of course not!" his father said. "I always do arithmetic in German."

"I wonder why you do that," Nicolaus said. "Do you suppose I will need to learn more languages? Are there other languages that educated people need to speak?"

"Yes, there certainly are," his father said, "but you will have to wait and see what other languages you will need. It is possible that you will want to travel to England, and that means you may need to learn to speak English, too. But for now, I think you should concentrate on improving your mastery of the four languages you already know."

By the time Nicolaus was 13 years old, the family had returned to Basel, and he soon entered the university there, carrying on his studies in both German and Latin. He passed the master's examination in law at the age of 16, and then completed the requirements for the licentiate in legal studies in 1715 at the age of 20. However, like his father and uncle, mathematics continued to be his real love.

Johann and Dorothea's second son, Daniel, was born in Groningen in 1700. At that time, Nicolaus was five years old and their sister Anna was two. After the family's return to Basel in 1705, Daniel, who was then five years old, entered school there. Both in Groningen and in Basel, Daniel learned mathematics from his brother Nicolaus. By this time the two boys had become very close, and, as their uncle Jacob and their father Johann had done when they were younger, they worked happily together. They made a happy pair as they explored that abstract subject together. Sometimes they worked in earnest silence; other times the roars of laughter as they attempted to capture one more difficult concept could be heard throughout the house. Fortunately, unlike Jacob and Johann a generation earlier, this warm relationship continued into their adult years.

"Anna," the children's mother Dorothea asked her older daughter one afternoon, "what are you children playing with so earnestly?"

"Oh, you don't need to worry about that, Mother," Anna explained. "It's just some games with mathematics. Nicolaus is teaching Daniel how to do it, and it's such fun!"

"I'm not so sure that is the best thing for a young lady to be learning," her mother said with some concern. "You might be better off spending that time practicing the piano."

"Oh, no, Mother, I'm sure it's all right," Anna assured her.

"You do seem to be having a good time together," her mother said. "I guess there's no harm in it, and you are making good progress at the piano too."

18
Johann Bernoulli's Son Daniel Grows Up

Many years later, Daniel wrote to his friend Christian Goldbach (a family friend whom Daniel would come to know well when he lived in St. Petersburg):

> My brother Nicolaus became a mathematician almost accidentally. Perhaps it was because it came so easily to him that he didn't realize what astounding progress he had made in mathematics. He wanted to instruct me in the calculus that our father had taught him as well as what he had figured out for himself, although when we began studying together I was only 11 years old. He used all his talent as he tried to teach me—his inadequate little brother. Only after I had learned it did he realize that in the process of teaching me he had truly mastered both the differential and integral calculus completely. In fact, I believe that his plan for us was that we should discover mathematics together—he never saw himself as my teacher, although that is precisely what he was. At that time he assumed that I was as accomplished a mathematician as he was! Foolish Nicolaus! I was nothing more than his ignorant apprentice who had been able to master a few small pieces of mathematics with a great deal of help from him.

When Johann, their father, realized that his oldest son had been teaching young Daniel mathematics, he decided to test Daniel's progress. "Daniel, come here, boy!"

"Yes, Sir?" Daniel replied uneasily.

"I understand your brother has been teaching you a bit of mathematics," Johann said. "See if you can solve this problem," which he wrote quickly on a piece of paper. Daniel was delighted to accept the challenge and happily took the problem to his room. He was pleased that his father was finally taking an interest in his efforts at mathematics. He quickly solved the problem, which was not difficult for him, and immediately brought it proudly back to his father.

"Father," Daniel said happily, "here is the solution to your problem."

"What took you so long?" Johann demanded. "You should have been able to solve that while you were standing here! I thought Nicolaus said you were good at mathematics. Bah! You'll never achieve anything important. What a pity!"

Daniel was devastated. He had always suspected that his father preferred Nicolaus, but he had dared to hope that this time he might have accomplished something that his father would find worthy. Alas, that was not to happen. Any mathematics that Daniel would do in Basel would be with help from his brother Nicolaus or on his own, always with less than no support from his father.

"Daniel," his mother Dorothea asked a little later when she found him quietly reading by himself, "what's the matter?"

"It's nothing, Mother," he replied.

"I heard your father saying something to you, and I think it made you unhappy," she persisted.

"Oh, no, Mother," he said, "You don't need to worry. I have been learning a little mathematics from Nicolaus, but I haven't gotten very far."

"But you like mathematics too, don't you?" his mother asked.

"Yes, I like it, but I'm nowhere near as smart as Nicolaus," Daniel said.

"I don't think that is the case at all," his mother said. "I think you are every bit as bright as your big brother. Since he is older, of course he is further advanced in mathematics, but I'm sure that with time you will learn it as well."

"I'm not so sure about that, Mother," Daniel said, "but I guess I'll try to do a little more mathematics with Nicolaus. Maybe I can do something."

"I hope you will, Daniel," his mother said. "I believe you will succeed."

A family friend, Condorcet, explained many years later that the family obtained the honor of Daniel's brilliant work in science in spite of itself. This was an honor that the family had no right to claim since the family (with the exception of his brother Nicolaus and possibly his mother) did nothing to help Daniel. What Daniel accomplished, he did because of his own passion and genius.

Daniel completed his first degree at the university in Basel in 1715 when he was 15 years old and completed his master's degree in 1716 when he was 16 years old. His father then picked out an attractive young woman from a good family, who could provide excellent ties to Basel's business community, to be Daniel's bride. Daniel was horrified—the marriage was clearly impossible. He was shy and would have felt overwhelmed and miserable in the company of this socially accomplished young woman.

Having failed in his attempt at arranging Daniel's marriage, Johann then instructed Daniel to prepare for a career in commerce, arranging an apprenticeship so he could learn the basics of business. Although Johann's father had never attempted to arrange plans for his marriage, he had certainly planned to set up Johann for a life in business. In the same way that Johann in his youth had rejected a life in business, his son Daniel also abhorred the plan. Johann failed to see that it was just as inappropriate for his own son Daniel as it had been for himself, regardless of whether it was a good way to earn a living. Daniel was a handsome young man with a charming, quiet wit when he was in the company of friends, but life in the business world would have been unbearable for him.

"Father, I have no interest in a career in business," Daniel protested.

"Did I ask you if that was what you wanted?" his father demanded.

"No," Daniel admitted.

"You will do as I say, young man!" Johann said.

Eventually Johann relented and allowed Daniel to study medicine instead, and nothing more was said about the suggested marriage. Daniel studied medicine first in Basel and then in Heidelberg, at that time part of the German Palatinate. His doctoral dissertation, which he completed in 1721 at the age of 21, concerned the mechanics of respiration from a mathematical viewpoint. When he had completed that, he applied unsuccessfully for a professorship in anatomy and botany at Basel and again the next year for the professorship in logic.

When it was clear that there was no position for Daniel at the university in Basel, his father Johann arranged for him to travel to Venice to study practical medicine with Pietro Antonio Michelotti, one of the most highly respected physicians in Europe, whose investigations into the way in which blood flows in the human body fascinated Daniel. Michelotti and Daniel worked most congenially together, sharing a love not only for medicine but also, unbeknownst to Daniel's father, for mathematics. Daniel had been so successful in helping Michelotti in a dispute with another Italian physician named Ricatti that Michelotti was delighted to help Daniel in his career, encouraging him to collaborate with him openly in his work both in the hospital and with his private patients.

Johann had also planned for Daniel to study with G. B. Morgagni in Padua, but serious illness forced Daniel to abandon that plan. After several weeks of feverish misery, Daniel finally limped back to health, exercising his mind with mathematics as he began to regain his strength.

Before Daniel left Venice, his mentor Michelotti and the Bernoullis' family friend Christian Goldbach helped Daniel publish his first book, *Exercitationes mathematicae* [*Mathematical Exercises*], a work of serious mathematics. That book allowed Daniel to launch his scientific career. Michelotti and Daniel celebrated happily when the book passed the censors' examination—a major hurdle in Italy at the time—and was actually printed.

With encouragement from Michelotti, Daniel also entered an essay in the Paris Prize competition—the equivalent of the Nobel Prize or the Fields Medal today—the ultimate competition for any scientist at the time. Although he was much too young and inexperienced to expect to win, it was still possible for a novice like Daniel, since every entry was made under a pseudonym, which kept the identity of the entrant sealed until all entries had been judged and allowed each entry to be judged on its merits. Daniel remained in contact with Michelotti for many years, discussing Daniel's mathematical and physical discoveries by letter. Each had profound respect for the other's abilities and insights. Their warm friendship may have been the closest that Daniel ever came to a constructive father–son relationship.

As a result of the publication of his book, Daniel was offered the position of president of a new scientific academy that was about to be established in Genoa, Italy, but he declined that offer. Daniel was uneasy with the political situation in Italy, given the tradition of censoring any scientific work that was seen as contrary to the teachings of the Catholic Church. Galileo had suffered from such censoring, and it appeared that little had changed since his time. Besides that, Daniel was eager to return to his homeland, where he knew his writings would never be subjected to anything like the Italian inquisition. As with many members of the Bernoulli family, Daniel's desire to live in Basel was strong as well.

19
Daniel Bernoulli, the Paris Prize, and the Longitude Problem

In the eighteenth century with its rapidly expanding international trade, technology that would allow a captain to determine the precise location of his ship at sea was an urgent challenge. Since navigation was imprecise at best, shipwrecks were a calamity that occurred far too often, as ships suddenly ran into rocky shores when they thought they were far from any land. Sailors had been able for many years to find their ship's latitude—how far north or south they were—by finding the angle of the sun at its highest point at local noon. However, determining their longitude—the distance east or west—was still a matter of guesswork. Not knowing the ship's longitude meant that the ship's location could be anywhere on a horizontal line stretching around the globe.

If someone could devise a method for knowing the precise time of day or night in a ship at sea, that would allow the sailor to calculate his longitude so that he then could pinpoint his location in the vast ocean. Although Huygens' pendulum clocks were fairly precise on land, they required a steady base and were useless on a ship tossing about for many weeks on the vast ocean. Unfortunately, developing a method to determine the exact time at sea was proving to be extremely difficult. An error of only one minute in 24 hours produced an error of 15 nautical miles or 15 minutes in latitude. During a journey of several months, those many minutes could be compounded to produce a fatal error. The Paris Prize hoped to encourage the scientists of Europe to solve the complex problem.

In 1725, 25-year-old Daniel Bernoulli learned that he had won the prestigious Paris Prize. He received 2500 *livres* [pounds] for his essay "On the Perfection of the Hourglass on a Ship at Sea," in which he described attaching an hourglass to a piece of metal floating in a bowl of mercury, thus minimizing the disturbance to the hourglass in a storm at sea. An hourglass seems to us in the twenty-first century like a primitive tool, but an hourglass can be carefully calibrated, allowing one grain of sand at a time to slip through the opening. Daniel's solution was one step in the desperate eighteenth century search for a method of determining longitude at sea. Although his hourglass was not part of the eventual solution, it was a step that appeared to offer some hope in the quest.

In 1747, at the age of 47, Daniel won the Paris Prize again with his submission of another work on the longitude challenge, sharing it this time with another entrant. This time Daniel's device was a method of controlling the vibrations of a combination of pendulums whose vibrations seemed to cancel each other out so that the result allowed him to power a reasonably precise clock that was only minimally affected by the tossing of the ship at sea. Daniel admitted that his results with the interacting pendulums were surprising, but they seemed to work, and the judges in Paris concurred. Since his solution required several intricate devices working together, however, it was still not the ideal solution. Nevertheless, it was another step in the eventual solution of the longitude problem.

Between 1730 and 1773, John Harrison—an English carpenter and clock maker with little education but with a happy combination of ingenuity and skill—tried another approach. He worked diligently at perfecting one chronometer (a precise clock) after another. His first devices were clocks made of a variety of woods, some of them self-lubricating—an important feature since the quality of lubricating oils available at the time was unreliable. Then he moved on to constructions with a combination of wood and some metal fittings, and then finally to metal alone. Harrison reluctantly had to abandon the use of wood as he perfected his use of fine metal gears and springs. His final result "H4" was a precise timepiece in the form of a watch that was both portable (it would fit in a man's pocket) and accurate to within 1/3 of a second in 24 hours. Since it could be depended on to lose (or gain) the same amount of time each day, that error could be corrected through careful calculation.

Whereas Daniel Bernoulli had written two essays as he worked on the longitude problem at the same time that he was working on many other challenges, Harrison devoted his entire working life of 60 years to the development of his devices. The final product was a wondrous and beautiful machine.

Harrison claimed the reward of £8,750—a phenomenal sum—from the British government. It was Harrison who found the desired solution to the problem on which Daniel had worked on and off, and it is Harrison who deserves the credit for saving untold lives and ships at sea. All four of his chronometers are now on display at the Royal Observatory in Greenwich, England. Alas, none of Daniel's models have survived.

20
Leonhard Euler

On April 15, 1707, Leonhard Euler was born in Basel, the Bernoulli's home city, to Margaretha Brucker Euler and Paul Euler. At the time, Leonhard's father was the chaplain at St. Jacob's parish, also in Basel. Leonhard's mother was the daughter of a distinguished vicar in the Reformed Church in Basel. Although Paul Euler's family had little education and even less money, Paul had enthusiastically studied mathematics with Jacob Bernoulli and soon became a friend of his professor's brother Johann, possibly sharing a room with him in Jacob's house for a time. When Paul Euler had completed his studies at the university, he qualified as a Protestant minister.

Church in Riehen where Paul Euler preached.

Soon after young Leonhard's first birthday, Paul was named pastor of the Reformed church in Riehen, a small town about an hour's walk outside of Basel, where he would preach every Sunday for the rest of his life. The residence that was provided for the pastor was cramped—only two rooms, of which one was designated as the pastor's study. The Eulers produced four children, of whom the oldest was Leonhard, followed by two daughters and then another son.

Leonhard Euler was an easy, cheerful child, who was curious about everything. His parents quickly recognized that Leonhard had a remarkably flexible and creative mind. His phenomenal memory was one feature of his makeup that carried him through situations in life that might have stymied anyone less capable. To his dying day, he was still able to recite the entire *Aeneid* of Virgil in Latin, even noting the location of any given line on its page in the version from which he had learned it 60 years earlier. From his student days he knew by heart the first six powers of the first 100 prime numbers—for example $7^2 = 49$, $7^3 = 343$, $7^4 = 2{,}401$, $7^5 = 16{,}807$, $7^6 = 117{,}649$—so that he could recognize them whenever he came upon them as he worked. Additionally, he knew all the standard formulas of trigonometry and calculus.

One day, when Leonhard was about four years old, his mother was distraught that her little boy was missing. Since he was a child who knew his way around the village, he would certainly not be lost. Could he have fallen down a well? His mother shuddered at the thought.

"Leonhard! Leonhard!" called his mother. "Where are you?"

Eventually she discovered him in the family's hen house, sitting on the floor, surrounded by chickens. "Leonhard," she gasped, "what are you doing?"

"I am making a chicken," Leonhard calmly explained.

"What?" she asked.

"When the hens make a chicken," Leonhard said, "they sit on an egg until it breaks, and that is when a baby chicken steps out. I have watched them, and I have decided that I want to do that too. See, here is my egg. It hasn't hatched yet, so I must continue to sit on it."

"Oh, dear me!" his mother exclaimed. "Leonhard, you are not a chicken."

"I'm sure it doesn't matter that I am not a chicken," Leonhard said confidently, "because I am sitting on the egg just the way the hens do."

"No, Leonhard," his mother said. "Only a hen can turn that egg into a chicken. You are a boy, and there are many things that you can do, but hatching a chicken is not one of them. A hen must sit on an egg for three long weeks before it hatches. You can't do that. Come along. Let's gather up the rest of the eggs and take them into the house. I need some eggs for dinner today, and I'd like to prepare boiled eggs for breakfast tomorrow."

"But what about my new chicken?" Leonhard asked.

"All eggs do not turn into chickens," she explained. "Besides, we have enough chickens without hatching these eggs. Remember, the reason we have a hen house is so that we can harvest the eggs, not the chickens."

"Are you sure that we don't need my new chicken?" Leonhard asked seriously.

"How many chickens do we have, Leonhard?" his mother asked.

"Let me see. We have one, two, three, four, five, six, seven, eight … Stop, hen, I'm trying to count you! … nine, ten, eleven, twelve—we have twelve of them," Leonhard said. "That was hard, Mother, because they kept moving while I was counting."

"You did well, Leonhard. We have twelve chickens, and twelve chickens are all we need," she said. "Come along."

"Yes, Mother," Leonhard said.

After they put the eggs on the table in the kitchen, his mother said, "Next we need to pick some cherries from the tree in the garden. Would you please bring the basket?"

"Oh, Mother!" Leonhard exclaimed when they reached the tree. "Look at all the beautiful cherries! Don't you think we should try one before we start picking them? What if they are not sweet and delicious?"

"What a funny child you are!" his mother said. "Yes, you may eat one now (and only one!), but we both know that it will be tasty. Our village of Riehen is famous for its wonderful cherries, and our tree is one of the best."

Plaque on church in Riehen: Leonhard Euler, 1707–1783, Mathematician, Physicist, Engineer, Astronomer, and Philosopher, spent his youth in Riehen. He was a great scholar and a kind man.

"Mmmmmm," Leonhard announced. "Could the next one be as good as this one? Don't you think we need to find out?"

"No, Leonhard," his mother said. "They will all be delicious, but if you eat them all now there will be none for dinner. What would your father say if I had to tell him, 'I'm sorry, Paul, we don't have any cherries for dinner today because Leonhard ate them all!'"

"All right, Mother. I won't eat any more. You and Father probably like the cherries just as much as I do. Oh, no! I can't reach them all!"

"That is why we are working together," his mother explained. "Since I am taller than you, I can reach the ones that are higher up. What you should do is to concentrate on the ones that you can reach. Later I will get someone to come out with a ladder and pick the ones that even I can't reach."

21
Leonhard Euler's Early Education

Paul Euler took the responsibility for educating young Leonhard in his first years, teaching him the basics of reading and writing and probably also the rudiments of the Latin language, but taking particular pains to teach him mathematics as well. For a mathematics textbook, he chose the *Coss*, the arithmetic and algebra textbook from which Jacob Bernoulli (his professor at the university) and Paul himself had first learned mathematics. The *Coss* would be a difficult book for an ordinary child of his age to begin his studies with, but Leonhard had two advantages: his father was an accomplished and enthusiastic scholar of mathematics, and Leonhard was a brilliant child. He had little trouble mastering even the most abstract concepts, and whenever he stumbled, his father easily answered his questions.

"Now Leonhard," his father began the first day of Leonhard's arithmetic studies, "we will begin on the first page of the *Coss*. Do you see that it explains that the book is written in two sections? The first section has 12 chapters, beginning with an introduction to the digits 1, 2, 3, 4, 5, 6, 7, 8, 9, and 0."

"Yes, Father," Leonhard said, "I already know how to count."

"All right," his father continued. "Do you know how the bigger numbers are written?"

"I think so," Leonhard said. "If we have just the digit two, it is simply two, but if we have two digits like 2–0, this time the two is worth twenty, not two. Is that right, Father?"

"Yes, that's right," his father said. "Can you read this enormous number for me on page two: 24375634567?"

"It's pretty hard to see with all those digits run together," Leonhard observed.

"How about if we put them in groups of three, like this: 24 375 634 567?" his father suggested. "That is actually the way we usually write big numbers."

"That's easy now," Leonhard said. "On the right we have five hundred sixty-seven. Would the next three digits give us six hundred thirty-four thousand?"

"Yes, they would," his father said. "What about the next three digits 3–7–5?"

"Would they be millions?" Leonhard asked. "Would that be 375 million?"

"Yes, it would," his father said, "but how do you know all this? I haven't taught you arithmetic before."

"I've seen you reading numbers, and they seemed pretty easy," Leonhard said. "Is it okay that I learned this before you taught me?"

"Of course it is, my boy!" his father said. "I wonder if you have already figured out how to add numbers. Look at this addition problem on page four:

$$\begin{array}{r} 16\,894 \\ 3\,245 \\ 6\,780 \end{array}$$

Do you know what to do?"

"Not exactly," Leonhard admitted. "Should I start at the right or at the left?"

"At the right," his father said.

"So that means that I should add 4 + 5 + 0 which gives me 9, so should I put the nine under the first column?"

"That's right," his father said.

"So then in the second column, I'll add 9 + 4 + 8 and that gives me 21," Leonhard said. "Is the one worth ten and the two worth 200?"

"Yes," his father said.

"So that means I should put the one in the second column, and carry the two over to the hundreds' column. Is that right?" Leonhard asked.

"Yes, my boy," his father said. "I can see that the *Coss* will not take you as long as I had expected. You are doing well."

That evening Paul Euler and his wife Margaretha sat talking after putting the children to bed. Paul had a question for his wife.

"Margaretha," he began, "when I worked with Leonhard on arithmetic today, I was surprised at how much arithmetic he already knows. You haven't been teaching it to him, have you?"

"Well," she admitted, "I taught him to count when he asked me about it (I remember one day perhaps a year ago when we were talking about how many chickens we have and he counted 12 of them), and he has certainly asked a lot of questions about everything, but I have never discussed arithmetic with him."

"That's what I thought," Paul said. "We must have a remarkably intelligent child, Margaretha. I'm thinking that he and I will be able to work through basic arithmetic and algebra within perhaps two years, and after that it will be time for him to go to school."

"But there is no satisfactory school in our village of Riehen," Margaretha observed.

"No, so I expect he will need to go to school in town—in Basel," Paul said.

"Perhaps he could stay with my mother," Margaretha suggested.

"Do you think she would be willing?" Paul asked.

"Yes, I think she would be delighted," Margaretha said. "She has been lonely since my father died, and she certainly has enough room for him. I think they would get along wonderfully. You know, Leonhard is a charming little boy. Shall I ask my mother?"

"Yes, I would like you to do that," Paul said. "I suspect Leonhard will be ready for the Latin school in town a year from next fall. You know, Margaretha, we are both quite intelligent, but I think our son is unusually bright."

"What did we do to deserve such a fine child?" Margaretha asked. "We are very lucky, Paul."

"Yes, Margaretha," Paul agreed. "He is a wonderful child. We are indeed very lucky."

22
Leonhard Euler Goes to the Latin School in Basel and Then on to the University

When he was seven years old, Leonhard walked with his father into Basel so that he could start school, having sent a trunk with his clothes ahead by carriage. It was an hour's walk, and Leonhard was impatient—he was eager to see his grandmother, whom he hadn't seen in more than a month, and now he was also going into the world and to school for the first time. His life was suddenly full of changes. After a brief visit with his grandmother, Leonhard and his father went to the Latin School to register for classes, which would begin the next day.

Paul discovered to his horror that the education at the school was not as complete as he had expected. Leonhard would be instructed in Latin and Greek and would learn to write in an elegant, fluent style, but it appeared that little else would be offered. Paul then investigated what he could do about this. Clearly, his son needed to continue his work in mathematics, the most beautiful subject that one could study at his age. With some help from his friend Professor Johann Bernoulli, who agreed that the Latin school's program was far from adequate, he contacted Johannes Burckhardt, a young student in theology who was one of Johann Bernoulli's students. With an excellent background in mathematics, Burckhardt was delighted to tutor Leonhard in that subject several days a week.

With those arrangements made, Paul kissed his son and his mother-in-law good-bye and returned to Riehen alone, reminding Leonhard that they would expect him home for the evening meal on Friday after his first week of school. He had no doubt that Leonhard could remember the way home, although in later years Leonhard described that hour-long walk as the most boring walk in the world. Young Leonhard had many ideas circulating in his small brain, and he was impatient to move ahead with them.

He did well in school, mastering penmanship, Latin and Greek, and continuing his study of mathematics with his tutor. He was a happy, curious boy, and he and his grandmother enjoyed one another's company. She provided him with a room to himself, which he had never had in their home in the parsonage in Riehen, and a table where he could do his lessons. After he finished his schoolwork in the evenings, they had spirited conversations on a wide variety of topics. Her charming grandson delighted his grandmother every day.

When Leonhard was 13 years old, he was ready to begin his studies at the university. He was the same age as most of the beginning students there, but he was undoubtedly better prepared than most. Paul talked with his friend Johann Bernoulli again, and the two of them agreed that young Euler should be enrolled in the elementary geometry and arithmetic courses, but Bernoulli soon determined that Leonhard did not need to sit through such basic courses in mathematics.

When Paul Euler asked Professor Bernoulli if he would be willing to tutor his son in mathematics, the great professor declined, realizing that in all likelihood the Eulers would not be able to pay the fees necessary for such tutoring. Instead, he suggested that Leonhard should study certain recommended mathematical texts on his own and then on Saturday afternoons Professor Bernoulli would be available to answer any questions Leonhard might have and to suggest the next book when the time came. When he was older, Leonhard Euler recommended this as the best way to learn mathematics. He had been forced to think critically by himself and to develop the mental discipline to pursue the concepts in a methodical way as he phrased his questions for Professor Bernoulli clearly and concisely.

At this time, Leonhard Euler became a close friend of Johann Bernoulli's two older sons, Nicolaus and Daniel. Euler often spent part of a Saturday afternoon with the Bernoulli boys either before or after his meeting with the great professor. The boys all shared a fascination with mathematics, and together they explored many concepts that were not officially part of the curriculum but that helped them all as they developed mathematically.

In 1720 at the age of 14, Leonhard Euler gave a speech in Latin to his fellow students at the university describing the beauty of algebra and geometry. He declared that nothing could compare with mathematics, and he urged his listeners to join him in this wonderful study. His delivery was flawless, and his professor of rhetoric was impressed.

Each week, Leonhard worked hard with the texts that Professor Bernoulli had suggested the previous Saturday. The work was difficult, and Leonhard would sometimes have to begin work on a topic more than once as he struggled to understand it. Sometimes he even had to give up and save that question for his mentor. However, he tried to take as few problems as possible to this very busy and important man. When Leonhard asked his first question on Saturday afternoon, he often discovered that Bernoulli's answer resolved his other questions as well. In his first year at the university, young Euler mastered first the differential and then the integral calculus. He was then ready to move on even further in his study of mathematics.

After two years of study at the university, Euler delivered another lecture, also in Latin, this time on temperance—practicing moderation in all things—for the completion of his first degree from the university. The following year he took part in a formal debate (also in Latin) on logic and the history of Roman law. In 1723 at the age of 16, he achieved his master's degree, delivering another speech (again in Latin) comparing Descartes' explanation of the set-up of

Leonhard Euler.

the universe with Newton's. In his speech, he presented Descartes' concept of the aether and explored whether it could explain the basic structure of the universe. Leonhard Euler aimed high and seemed to be able to succeed at whatever he chose to do.

Next, Leonhard enrolled in the school of theology, because his father had determined that Leonhard should follow in his footsteps and become a minister too, in spite of the fact that, like Jacob Bernoulli before him, Leonhard Euler was far more interested in mathematics than in theology. Professor Bernoulli decided that Leonhard was so talented in mathematics that he should have a talk with his friend Paul Euler—such a talk as no one had ever dared to attempt with his own father or with Johann himself concerning his own sons. The Bernoulli reaction to another point of view concerning a son's career was always the same: no!

"Euler," Johann Bernoulli began, "I have truly enjoyed working with your talented son in mathematics."

"He seems to be a very bright young man," Paul Euler agreed. "I'm very pleased with his progress."

"You should be! Now, Euler, I realize that you are eager for your son to have a career in the Church," Bernoulli continued, "but I think that would be a waste of his enormous talent. Couldn't you allow him to concentrate on mathematics instead?"

"Do you think that would be better for him?" Paul asked in astonishment. "He has done well in all his studies, and I think he would be a distinguished scholar of theology."

"Yes, I'm sure that your son has excelled in all subjects and would in all likelihood do the same in theology," Bernoulli said. "However, he is without exception the brightest mathematics student that I have ever taught."

"My goodness!" Paul said. "Coming from you, that is quite a statement! I already knew that he was unusually intelligent, but I presumed that as his father I am prejudiced. If you don't mind, I'd like to discuss this with Leonhard and his mother before I make any decision. It is his life, and I don't want to make his decisions for him. I would certainly love for him to have a career in the Church, but perhaps that is not what is best for him. Thank you for bringing this to my attention, Bernoulli. I realized long ago that my son is enchanted with mathematics—so am I, as you probably know. I'll have him talk with you after we have discussed it."

"That is an excellent idea," Bernoulli said. "I wish you well in this, Euler."

"Thank you so much," Paul Euler said as the great professor left.

The next Saturday, Leonhard walked home to Riehen as usual. When he got there, Paul Euler and his son sat down for a chat.

"Leonhard," his father began, "Professor Bernoulli has talked to me about your future studies and your career. He is recommending that you specialize in mathematics rather than theology."

"Really? He came to talk with you about me?" Leonhard asked.

"Yes, he did," Paul replied. "I was surprised as well."

"Well, Father, it is true that I am fascinated with mathematics," Leonhard agreed, "but you have always wanted me to become a minister. I like to follow your directions."

"Yes, you have always done that," Paul said. "You have always been a good and obedient child, but now you are becoming a young man, and it is your life. I believe your opinion should matter as well. What do you want? Do you want to become a mathematician instead of having a career in the Church?"

"I certainly wouldn't mind studying theology," Leonhard said, "but I guess my real passion is for mathematics. I can think of nothing more wonderful than becoming a mathematician. What fun that would be!"

"That's what Professor Bernoulli thought," Paul said, "so that is clearly what you should do."

"But I don't want to disappoint you, Father," Leonhard said.

"Nothing you could do would disappoint me, Leonhard," his father said. "I expect you will be a devout, God-fearing man, but you don't have to be a theologian to do that. Let's talk with your mother about this before we settle down for our evening meal. If she agrees, would you please go talk with Professor Bernoulli and tell him that we have discussed this and that you will take his advice and concentrate on mathematics instead of theology?"

"Thank you, Father!" Leonhard said. "That makes me the happiest young man in all of Riehen! Let's talk with Mother, and then I will talk with the professor first thing on Monday." Leonhard strode over to the table and helped his mother put out the last things for supper.

"Yes!" Paul Euler said quietly to himself. "No one could ask for a finer son! And for him to be a genius on top of that is more than Margaretha and I ever hoped for, and it is certainly more than we deserve. Thank you, God, and may we always find the guidance we need to help him as he makes his way in the world."

23
Daniel and Nicolaus Bernoulli Receive a Call to the Academy at St. Petersburg

In 1724, when Daniel Bernoulli at the age of 24 published his book *Exercitationes mathematicae* in Venice, his father was surprised that his second son was making a name for himself in mathematics—not in medicine as his father had planned. In an attempt to mollify his father, Daniel identified himself in the introduction to his book as "Daniel Bernoulli, Johann's son," a practice he continued in all his scientific writings, even long after he had established himself as a distinguished professor and scholar. His father never acknowledged Daniel's gesture, either then or later. Nevertheless, Daniel humbly continued to portray himself as a lesser scholar than his father, the respected prince of mathematicians in Europe.

In his book, Daniel began by reviewing the literature on probability, including his uncle Jacob's complete study of the subject, *Ars Conjectandi*, which had been published 11 years earlier. With that background for his reader, Daniel then explored probability as it applied to a card game called *faro*, which was popular at the time. Daniel concluded that the game favors the dealer, but only very slightly, meaning that as long as the game was honestly played, any player had a good chance to win. Daniel might not have been surprised to learn that in the nineteenth century, *faro* became the most popular game in the American West, probably because it is a fair game that is easy to learn and quick to play. Cowboys enjoyed playing a game of chance with their pals after a long day on the trail, but only if they could win often enough to keep money in their pockets.

The second part of Daniel's book dealt with the movement of liquids, specifically how they flow from openings in a container, drawing on Daniel's interest as a physician in the flow of blood through the human body and hinting at his future interest in the field that he would name fluid dynamics. The third and fourth parts of the book were pure mathematics, concerned with differential equations and geometry.

Daniel's book was so well received that when he returned to Basel in 1725, having turned down the position at Padua, two surprises were waiting for him: Not only had he won the Paris Prize for his essay on developing a more accurate hourglass that would work on a ship at sea—a remarkable accomplishment for a budding scientist—but he was also invited to join the newly formed St. Petersburg Academy in Russia as an accomplished professor. St. Petersburg would have preferred to get the services of his father, but his father had no interest in leaving Basel. Daniel was an acceptable second choice for the fledgling academy.

Daniel's father Johann wrote to a friend, celebrating his triple happiness over his son Daniel's successes, mentioning "the return of my son [after his serious illness], the Paris Prize that Daniel has won, and his appointment to the academy at St. Petersburg." Considering the way that Johann would later treat Daniel, it is unclear how heartfelt these thoughts were. A cynical view might suggest that he was simply celebrating his relief at no longer being responsible for the support of his second son. On the other hand, a Bernoulli who was succeeding so admirably in mathematics and the sciences couldn't do the family name any harm as long as his accomplishments posed no threat to his father.

Johann wrote at the time to another friend that Daniel had been "offered an annual salary of 600 Rubles, free housing, and a sufficient quantity of firewood and candles..., an honorable offer to a young man who is only 25 years old." For his part, Daniel was reluctant to go so far from home—and he may have thought that the brutal northern climate would require a prodigious amount of firewood and candles if one wanted merely to survive there. However, when his 30-year-old brother Nicolaus (who was occupying the chair of law at the university in Bern, Switzerland) heard about Daniel's offer, he enthusiastically volunteered to accompany Daniel to Russia, where together they could pursue mathematics, the subject that they both loved more than any other. Nicolaus had no qualms about abandoning the career that his father had chosen for him in law, not a favorite subject for anyone named Bernoulli, regardless of the wishes of his father.

Daniel then wrote a letter to Christian Goldbach, who had recently become the permanent secretary to the newly formed Academy at St. Petersburg, observing that "there is nothing so vast as the study of mathematics," and that the institute at St. Petersburg looked as if it would be a wonderful place to explore that exciting subject.

Christian Goldbach.

Then Daniel asked Goldbach if he could use his influence to arrange for a place for his brother Nicolaus in St. Petersburg as well. Daniel explained that if Goldbach could accomplish that, he would "have the merit of not separating two brothers whose friendship was the closest that the world had ever seen."

Since Goldbach and Nicolaus had become friends when they both were in Venice a few years earlier and Goldbach had since come to know Daniel as well, he was pleased to help. Already aware of the vast talents of the Bernoulli family as well as the strife between Daniel's father Johann and his uncle Jacob, Goldbach was happy to use his influence to arrange for these Bernoulli brothers to become happily cooperating scholars at the St. Petersburg Academy of Sciences. Jacob and Johann's disputes had been clear for all to read, as they had sparred openly on the pages of the *Acta Eruditorum*.

In early June, Daniel wrote a second letter to Goldbach, thanking him for his help in arranging for Nicolaus to be invited to St. Petersburg:

> Since I know that you took part in these arrangements, I should tell you that I have accepted the chair of mechanics [at St. Petersburg].... They were so generous as to offer me an annual salary of 800 Rubles plus extra money to pay for the journey. I promise you, *Monsieur*, that this was a very difficult decision for me. Should I accept it? In the end, I hope that I will not be blamed too much for my delay, having never complained about the terms that were first offered—being uncertain only about how and where I wanted to pursue my career. I found the initial offer of 600 Rubles sufficiently generous, and I am not so presumptuous as to have mercenary thoughts. My concerns had to do with the path of my career, not money.

Since there was no appropriate position in mathematics or physics for either Daniel or Nicolaus in Switzerland, their move to St. Petersburg was wise.

Daniel commented to his brother soon after they arrived in St. Petersburg, in time for the official opening of the academy there on November 13, 1725, "You know, St. Petersburg is not as bad as I feared. The laboratory facilities are magnificent—I never dreamed of such a work environment! It's cold and dark, and I know it will get colder and darker before the sun finally returns in the spring, but the group of scientists assembled here is pretty impressive! I think we can do some good work here. Don't you find the intellectual atmosphere truly exciting?"

"Oh yes, Daniel," Nicolaus said, "I wasn't worried about that. This is a vast improvement over my career in the law, which I have to admit I found very, very dull. Also, what fun it is for us to work together again! By the way, aren't you impressed with our colleague Jacob Hermann? He is an impressive mathematician and a warm and wonderful man. I like him very much."

"Yes, and the fact that he speaks German with a beautiful Swiss accent makes me like him even more," Daniel admitted.

"You realize that Hermann studied mathematics at Basel with Uncle Jacob, don't you?" Nicolaus asked.

"Yes, although Father would not necessarily see that as a plus," Daniel said.

"Whatever Father says," Nicolaus said, "Uncle Jacob must have been a superb mathematician, and I hear he was a brilliant teacher as well. I don't know what it is about our father. He seems to worry a lot about his status in his relations with other mathematicians. Deep in my heart, I think Father may have been the one who caused much of the unpleasantness with Uncle Jacob."

"I can believe that!" Daniel said. "I wish I had known Uncle Jacob. He must have been an impressive mathematician. Do you remember him?"

"No, I was only a baby when we left Basel for Groningen, and by the time we returned he had already died. If I ever met him, it was only to be admired as a small baby," Nicolaus said.

"But still," Daniel said, "we have to admit that our father also is an incredibly gifted mathematician and teacher."

"That is true," Nicolaus said. "But you know, Daniel, as I see it, Father hasn't always been very nice to you either! Like Uncle Jacob, I'm afraid you have often been at the receiving end of his venom."

"Well," Daniel admitted, "I have to admit that I have never felt like his favorite son although there is no doubt that you are a better mathematician than I am."

"You are wrong there, Daniel," Nicolaus said. "Your talent in mathematics is truly phenomenal, but this argument isn't going anywhere. We certainly work well together, and let's just enjoy that. I think we are very lucky, and I firmly believe that the scientific world will benefit from our work. I'm so glad Goldbach was willing to intervene on our behalf. You were clever to ask him to do that!"

Before their departure for St. Petersburg, both Nicolaus and Daniel had become friends of the brilliant Leonhard Euler, who was seven years younger than Daniel and was their father's star student. The three of them decided even then that a position in St. Petersburg would be ideal for Euler, since Basel had little to offer Euler as well. They knew their father was slow to express his approval of any student in mathematics, and yet he was unstinting in his praise of young Euler. That must mean that Euler's talents were truly impressive! Furthermore, there was no question that Euler would be a most congenial colleague. Before Daniel and Nicolaus set out for St. Petersburg they promised Euler, who had almost completed his studies with their father, that they would work to arrange a chair for him in St. Petersburg. The brothers accordingly made their initial steps on Euler's behalf almost as soon as they arrived in St. Petersburg, making sure that Goldbach in particular understood what a fine scientist their young friend was.

As their first winter in St. Petersburg wore on, the two brothers were happily working on several difficult questions in mechanics and mathematics, but soon Nicolaus, who was only 30 years old, showed signs of severe, chronic pain in his abdomen.

"Nicolaus," Daniel looked at his brother with concern, "what's the matter? You look as if you are not feeling well."

"No," Nicolaus admitted, "this pain in my stomach just won't go away. You studied medicine. What do you think it might be?"

"I wish I could give you an easy answer," Daniel said. "I wonder if you have an ulcer. Maybe you should try to eat more bland food. We had spicy goulash for dinner yesterday. Do you suppose that affected you?"

"Possibly, but it certainly did taste good!" Nicolaus said.

"Why don't you try restricting yourself to puddings and cheeses, and avoiding anything spicy or acidic," Daniel suggested. "I could ask the cook to provide bland foods for you."

"I suppose that couldn't hurt," Nicolaus said. "Our research is hard enough without a constant stomach ache."

Unfortunately, less than a year after they arrived in St. Petersburg, Nicolaus died at the age of 31 of an intestinal infection—perhaps due to a stomach ulcer. Daniel was devastated. He had done his best to help, but his medical knowledge, which was as complete as anyone had at the time, was simply not adequate. His dearest brother and best friend, who had been willing to travel with him to the brutal climate of St. Petersburg so that they could work together there, had provided the major reason that Daniel had been willing to go to Russia. Without Nicolaus, the harsh environment in St. Petersburg was unbearable for Daniel. His only consolation was the prospect of the arrival of his young friend Euler. Daniel never had been happy in St. Petersburg—it was too dark and too cold—and he yearned more than ever for a return to his home in more temperate Basel.

24
The Academy of Sciences at St. Petersburg

Long before the Bernoulli brothers arrived in St. Petersburg, Peter the Great of Russia had had plans for his realm that were, like himself, bigger than life. At a height of almost seven feet, he towered over everyone else, and he planned for his new capital to be equally impressive. When he began plans for St. Petersburg, his fantastic "Venice of the North," he hired a bevy of internationally famous architects to plan his city at the swampy mouth of the Neva River, according to a spectacular geometric plan to be carried out through the work of thousands of slave laborers. Although Russia was still a feudal society, Peter the Great had high hopes for a more enlightened kingdom in the future. However, his ambitious plans could not be carried out without a large labor force, and for this project the Tsar's slaves conveniently provided such a group of workers. Social and economic equality, wonderful ideals that he would love to foster, would have to wait. By the time Euler arrived, St. Petersburg had already become a beautiful city, although the political climate was far from stable.

When Peter the Great had visited the French Academy in Paris several years earlier, he dreamed of the new capital that he would found. Quite naturally, its crowning jewel would be an academy of science on a par with the one in Paris. The great polymath Leibniz, who had met with Peter the Great three times between 1711 and 1716, had made ambitious plans for such new academies, hoping to found them in capitals throughout Europe. Peter decided that his city and his academy would be their finest example. Scholars there would find a utopia where they could explore science freely and happily, bringing fame to his Russia in the process. Peter assumed that, following Leibniz's plan, his resident scholars would also occasionally use their skills to solve whatever practical problems the monarch might refer to them, but most of their energies would be devoted to their own scholarship. The Tsar found that a charming idea.

As he had made his plans, Peter had hoped to establish the faculty of his academy with local Russian talent to serve as professors. However, since there were so few accomplished Russian scholars at the time, he decided it would be more efficient to import established scholars from outside of Russia, believing that those scholars would give his academy instant prestige. Since one of the functions of the academy was to provide a secondary school and university for the education of the next generation of Russia's leaders, the end result would be an impressive growth in the ranks of Russia's own scholars, so that in the future the professors could be Russians.

In founding his academy in 1724, Peter had tried to bring the great mathematician Johann Bernoulli, father of Daniel and Nicolaus, to St. Petersburg instead of his two sons. Johann Bernoulli, however, wasn't tempted. While Basel was no intellectual capital of Europe, Johann was already recognized as the foremost mathematician in Europe, and he felt no need for a more invigorating scientific environment. He had never needed the stimulation of other scholars, whom he was much more likely to view as unwelcome competition rather than as congenial colleagues. Johann had countered the Tsar's invitation by suggesting that his second son would be perfectly adequate for the Tsar's needs. When the Tsar and later the Tsarina actually were able to attract not just one but then two of Johann's sons, they felt doubly lucky.

Up to this time, only the Prussians had actually followed through on Leibniz's plans, forming the Prussian Academy of Sciences in Berlin with Leibniz as its president. However, the academy in Berlin was not as successful as Leibniz had hoped. Since Leibniz had so many different projects demanding his attention, he had little time for actually administering the academy. He was trying valiantly to finish his history of the royal house of Brunswick, which in fact he never completed, and he was spending more and more of his time fighting in the priority dispute with Newton over the discovery of the calculus. Thus the Prussian Academy had already lost its vitality—Eric Temple Bell described it as "dying of brainlessness" at the time—and would offer St. Petersburg little competition at least for another ten years. There was no doubt that the Paris Academy was in a totally different league altogether, but Peter could hope for greater glory in future years.

When Peter the Great died in 1725 at the age of 52 from urinary tract and bladder ailments, Peter's widow and successor, the Tsarina Catherine I, carried out his plans as far as she could, but with her death only two years later at the age of 40 in 1727, nothing was assured. When she died, the 12-year-old Tsar Peter II began his reign, just as Euler was arriving in St. Petersburg. The resulting struggle for power among the Boy Tsar's relatives disrupted all of Russia, since as a child there was no doubt that he was incapable of ruling alone.

As a result, Daniel Bernoulli found himself welcoming Euler to a precarious academy in a headless country with an uncertain future.

25
Euler Begins His Career and Moves to St. Petersburg

In Basel in 1723, 16-year-old Leonhard Euler had completed his master's degree in philosophy at the university. Having learned all the mathematics that was offered by his mentor Johann Bernoulli, the most important mathematician in Europe, Euler was ready to pursue mathematics on his own. Unfortunately, the university at the time offered no formal degree program in mathematics. In 1726, he published his first essay in the *Acta Eruditorum*. It was three pages long, and in it he explored the construction of isochronal curves within a medium that presents friction. The study of this curve, which had been begun by Huygens and continued by Jacob and Johann Bernoulli, originally concerned the curve followed by the bob of a pendulum clock. Euler's excellent essay on the subject, written in Latin, as was the custom, carried the study of the isochrone further and deeper.

In 1727, with encouragement from his mentor Johann Bernoulli, Euler submitted to the Paris Prize competition an entry exploring the best way to locate masts on a sailing ship. Novice that he was, he was awarded second place, after Pierre Bouguer, a professor of hydrology in Paris who had been the expert in the field for many years, and for whom the challenge had actually been written—Bouguer was supposed to take first place, and he did. Euler's accomplishment, as a 20-year-old scholar who had never seen a sea-going sailing ship nor even the sea itself, was impressive. He admitted in his introductory remarks that he had not submitted his plans to a practical test, which he couldn't have done even if he wanted to, living in the tiny land-locked Helvetian Confederation, far from the sea. However, he stated that trials were unnecessary, since he knew that his work was correct because it followed the well-established principles of mechanics. The judges at the Paris Academy agreed.

Also in 1727, Euler completed his *Dissertation on the Theory of Sound*, which is often considered his PhD dissertation, and with its publication he competed for the chair of physics at the university at Basel. In fact, he was not seriously considered for the position, mainly because he was so young. While he was still less than 20 years old, several other contenders were much older and better established in their careers. Although his competitors may have seen young Euler's application as laughable, his mentor Johann Bernoulli fought hard for him, saying that no other candidate could possibly be half as well qualified as Euler. He was undoubtedly right, but the choice wasn't his to make.

In fact, it is probably best for Euler that he did not win the professorship in 1727. Basel never had another chance to attract the genius Euler to return to his home university, since after this Euler would pursue his career internationally, not locally. In 1748, when his mentor Johann Bernoulli died, the university immediately wrote to then 41-year-old Euler, inviting him to come and replace his mentor, probably never dreaming that he would turn them down. By this time, however, Euler had no need for Basel's comfortable environment. He was already recognized as the most important mathematician in the world. The reason that Basel had enjoyed respect in the academic world of mathematics for so many years was because the brothers Jacob and Johann Bernoulli simply preferred to live there. Once they were gone, the major events of mathematics took place elsewhere.

Since young Euler wanted to make his mark as a scholar, he was far better off in the midst of a thriving institute, working with a group of world-class scholars in an academy with its own scholarly journal in which he could and did publish his discoveries. In the end, he outdid every other scholar in Europe of his time, creating and writing more original mathematics than any other single scholar has ever done within one lifetime.

When he had left Basel in 1727, Euler had been offered a lowly position as adjunct at the St. Petersburg Academy of Sciences. His friends Daniel and Nicolaus Bernoulli in St. Petersburg, the Bernoullis' father Johann (Euler's mentor) in Basel, and their friend Christian Goldbach had all worked diligently to make it possible for Euler to move to a position in St. Petersburg.

In the letter that Daniel wrote him accompanying the offer, Daniel wrote,

> You are awaited with great impatience; please come as quickly as possible. However, if you find the harsh winter of St. Petersburg daunting and you decide to wait until spring, I recommend to you, my friend, that in the meantime you brush up on your knowledge of human anatomy.... Also, please don't forget to send some pieces of your work to the academy so they can see for themselves that my recommendation is just.

Daniel admitted that it looked as if the initial position for Euler at the academy would be in the field of medicine rather than mathematics.

"I'll be glad to do that," Euler replied in his return letter. "Medicine is a field of which I know practically nothing, but I'm sure a little knowledge of anatomy wouldn't hurt."

Daniel wrote back:

> I'm glad you are willing to be flexible. As happens so often in academic environments, you should be able to move into mathematics over time. I have to admit, though, that I can hardly wait for you to arrive. And, don't forget to bring lots of warm clothes. This is a brutal climate. All the firewood in the world will never keep you warm enough here.

Acquiring the talents of Leonhard Euler turned out to be a magnificent move for the Academy at St. Petersburg—far better than anyone in St. Petersburg or perhaps even his mentor back in Basel could have predicted. Taking Daniel's advice, Euler waited to make the trip to St. Petersburg until after the harsh winter had passed. Following a few intense months studying medicine and anatomy, Euler set out on April 5, 1727, from Basel, going by riverboat down the Rhine River as far as Mainz and then up the Main River the short distance to the Free Imperial City of Frankfurt. From there he went by coach to Hamburg, and then to Lübeck on the Baltic coast. In that busy port, he boarded one of the first sailing ships that he had ever seen, traveling in it along Lübeck Bay and finally across the Baltic Sea, and arriving in St. Petersburg six weeks later in May of that year. Although the sea passage was rough at times, and Euler discovered for himself the horrors of sea-sickness, the young scientist was eager to begin his career.

When he arrived in St. Petersburg, Euler could see immediately that Russia under the new child tsar, Tsar Peter II, was in chaos, and, ever the pragmatist, Euler began exploring the possibility of a career in the Russian navy as a medical officer, since he now possessed some marketable medical skills. However, with the help of Daniel Bernoulli and Goldbach, Euler was able to turn down his offer from the navy and moved in with his friend Bernoulli at the Academy to begin his academic career with enthusiasm. Without apologies, Euler was somehow able to begin in the departments of mechanics and mathematics, rather than medicine.

In spite of the uncertain political climate, Euler found the Russian weather bright and pleasant that spring, although he was under no illusions about the harsh climate that he would face when winter set in. As far north as he was, in June the sun shone until 11 P.M. every evening and rose again only a few hours later, providing more productive hours of daylight than even Euler could use to full advantage. However, Daniel did not hesitate to remind his friend Euler of the inhospitable climate that awaited him in that northern city during the long dark winter to come.

"Euler, do you realize that there are no major cities anywhere in the world that are substantially farther north than St. Petersburg at 59° north?" Daniel asked his friend.

"Yes, I looked at the map before I started," Euler agreed.

"You know, the cities of Stockholm, Oslo, and Helsinki, all in Scandinavia, are at about the same latitude—either 59° or 60°," Daniel explained. "The only city farther north is Reykjavik, Iceland, at 64°, but it's a real stretch to call it a city—it is nothing more than a frontier outpost."

"I'm glad we didn't choose to go there," Euler admitted, "although I wonder if we would really notice a difference in moving only four or five degrees farther north."

"No sensible European would ever choose to live in Iceland," Daniel said. "Then again, why would any sensible European be so foolish as to live as far north as we are in St. Petersburg. Brrrrrrrr."

"Oh, come on, Bernoulli," Euler remonstrated with his friend. "Do you find that the climate interferes with your research or your writing?"

"No, I suppose not," Daniel admitted, "but I prefer living in Basel. I predict that you'll be singing a different tune when it is still dark and frigid in March, at a time when the sun is already high in the sky in Basel. You know, when Peter the Great established this city of St. Petersburg, he called it the Venice of the North. He must have had a magnificent sense of humor! Have you ever seen Venice, Euler?"

"No," Euler admitted. "This is the first time I've ever traveled anywhere outside of the Helvetian Confederation."

"Well, let me tell you," Daniel said. "I spent a few years in Venice, and this wasteland looks nothing like it! Peter the Great had ambitious plans, but there was no way he could move this city to Italy. I have to admit, though, that I'm glad Peter decided not to import the Italian censors as well. We can at least be grateful for that."

"That's right," Euler said, "you had to deal with them in order to publish your book, didn't you?"

"That's right. I didn't like it one bit," Daniel admitted. "It passed, but I think it was close. I resented the inquisition!"

One of Euler's decisions on arriving in Russia was to learn the Russian language as quickly as possible. Unlike Daniel Bernoulli, who constantly dreamed of returning to a life in Basel, Euler was a pragmatist who was determined to make his career in St. Petersburg. Learning the Russian language was an obvious and necessary first step. Being unable to communicate with the people around him would have been a handicap he didn't need, and with his phenomenal memory and quick wit he became fluent in Russian in just a few short months. This meant that, unlike most of his colleagues at the Academy, when he taught at the Academy high school and university, he was able to communicate easily with his students in their own language.

As time went on, the textbooks that he wrote for Russian schools were classics that would continue to be used for many years. From the beginning of his career, Euler wrote explanations of difficult scientific principles clearly enough that any serious reader could understand them, no matter whether he was writing in French, German or Russian for a general audience, or in Latin for the greater scientific community. Deliberately complicated prose was anathema to Euler; the purpose of his writing was to communicate, and to him clarity was paramount.

Euler and Daniel Bernoulli shared Daniel's comfortable, nicely furnished apartment at the academy for six years, enjoying one another's company and creating fascinating mathematics and physics together during all their waking hours. Daniel was not disappointed in Euler, the successor to his brother Nicolaus, who had come to join him in that desolate northern city. If he had been more like his father Johann, Daniel might have seen his brilliant young friend as unwelcome competition, but jealousy was not part of Daniel's makeup. The two young scientists collaborated happily, each respecting the other's accomplishments, and easily providing a different perspective as they struggled together on one difficult problem after another.

However, in spite of their congenial working relationship, Daniel still loathed the harsh Russian climate, and he continued to work toward his eventual return to his beloved Basel. During Euler's first few years in St. Petersburg, Daniel applied frantically for three different openings on the faculty of the university at Basel, but each time without success.

When the Boy Tsar died of smallpox three years after Euler had arrived, just before the Tsar's planned marriage, Anna Ivanovna ascended the Russian throne, and the atmosphere at the academy improved somewhat. Still, this was not the environment that Peter the Great had envisioned for his academy. The man who was effectively in charge of the Academy through all these political changes was an arbitrary, difficult man named Johann Daniel Schumacher (1690–1761). His official position was librarian and chancellor, and both Bernoullis as well as Euler found him irksome to deal with. In the turmoil after the Tsarina's death, Schumacher had skillfully seized arbitrary power, and for many years no one had the authority or finesse necessary to rein him in.

Schumacher's petty rules and unreasonable demands were a continual interruption to Euler's and Bernoulli's research. Not surprisingly, Schumacher's unpleasant methods were responsible for the departure of several capable scientists from the Academy. Euler, a mild and even-tempered man, suffered greatly but quietly under Schumacher's difficult personality for many years, choosing to keep his head down and continue to work rather than to fight. Euler was a survivor, whose religious background encouraged him to be humble and accommodating, and to trust in God that he would survive whatever problems he encountered. Daniel, who found it difficult to take that approach, was still unhappy, and he was determined to do something about it.

Euler's first position at the academy at St. Petersburg was only as an adjunct in mathematics, a lowly job with little pay and little respect. His initial salary, which was negotiated from Basel at 200 Rubles per year, had been increased to 300 before he actually made the trip, and, like the Bernoullis', Euler's offer included free firewood and candles. In addition he was awarded 100 Rubles to cover his travel expenses.

When Euler arrived, Daniel already occupied the position of professor of mechanics. Three years later in 1730, the Swiss mathematician Jacob Hermann, who was a distant relative of Euler, returned to Basel, and Daniel was able to move over to the more desirable chair of mathematics. That made room for Euler to be named professor of mechanics. Since Hermann, who also loathed Schumacher, had been extremely helpful to both Daniel and Euler in their first years in St. Petersburg, the two young men were sad to see him depart. However, since Jacob Hermann had commanded a generous salary, with his departure there was more money available for a promising, young scholar like Euler. As a result, Euler's salary was increased from 300 to 400 Rubles with an additional 60 Rubles for firewood. Later, with a new four-year contract, his salary was further increased to 600 Rubles. With his new title of professor and the accompanying salary, he also enjoyed greater respect from his peers.

26
Daniel Bernoulli and Leonhard Euler: An Active Scientific Partnership

When Leonhard Euler arrived in St. Petersburg, Daniel already had a modern working laboratory.

"Bernoulli," Leonhard Euler said to his friend, "please tell me what you are doing with this bizarre contraption."

"I've been experimenting with the flow of liquids," Daniel explained. "I got the academy's mechanic to construct for me this two-part watertight container made of copper. As you can see, the upper part has a diameter that has twice the diameter of the lower part. When I tilt the apparatus at this angle, I observe the water as it flows down. I also directed the mechanic to make me a bigger container to hold the whole apparatus so that I could catch the outflow as I work."

"Very interesting," Euler said. "I have to tell you that your laboratory is magnificent! You seem to have everything that a scientist could want. Do I assume that as you need new things, you have only to ask and they are provided?"

"Yes," Daniel replied. "It is really quite remarkable."

"But can you tell me exactly what you are looking for with your apparatus?" Euler asked his friend.

"First, I wondered how the speed of water as it flows downhill is affected by a change in the diameter of the channel," Daniel explained.

Bagaduce River in Maine, showing increased speed as the channel narrows.

"And what seems to happen?" Euler asked.

"It definitely goes faster, but unfortunately I haven't yet figured out how to measure exactly how much faster," Daniel said. "When I drop a wood shaving in the water, look at what happens." Both men paused to watch.

"Hmmm," Euler said. "It certainly goes much faster when the channel narrows."

"Yes," Daniel said. "Do you have any thoughts on why that happens?

"Not off the top of my head," Euler said.

"Well, I'm thinking that it has to do with the general flow of the water due to the force of gravity," Daniel said. "In the big vat, I believe the water should move at a standard speed, whatever that speed is. I'm thinking that when the channel is narrowed, gravity still wants to move the same quantity of water since gravity is a uniform force (Thank you, Mr. Newton!), but it has to speed up in order to allow that to happen through the narrower channel."

"You may well be right," Euler said, "but I'm afraid it's something I've never thought about before, so I really don't have a strong opinion on it."

"Now what I'm wondering about is how I can measure the speed of the flow precisely," Daniel said, "other than by watching how fast the wood shaving moves. I've tried timing it carefully with my clock, and I think that may allow me to get some good data."

"What are you thinking of doing with that data?" Euler asked.

"What I'm beginning to wonder about," Daniel said, "is whether there might be a change in pressure on the walls of the channel as the speed changes."

"And?" Euler asked.

"I've asked the glass blower to make me some narrow out-flow tubes that I can insert in the walls of the big container and also in the wall where the channel is narrower," Daniel said. "What do you think?"

"I hope you have a good mop!" Euler said with a big grin.

"Oh, I do!" Daniel said. "Look over there. I also have a bucket! I've already had many spills."

"I think you might be on to something important," Euler said. "I'll be glad to see what happens. Newton did a lot of work on the ways that solids behave, didn't he? I guess he never thought much about liquids."

"He may have thought about them," Daniel said, "but he never wrote about them as far as I know, although I haven't seen as many of his writings as I would like to. They were certainly never available in our house as I was growing up."

"No," Euler said, "your father seems to have little respect for Newton. Although I have great respect for your father, I think he is wrong about that. I think Newton was an amazing genius."

"I think it would be accurate to say that my father vehemently hates Newton and anything that Newton did," Daniel said, "and I agree with you: my father is wrong. I think it's possible that one of the things my father hates about my work is that I have written positively about some of Newton's ideas." In the centuries since Daniel did his work in St. Petersburg, it has been agreed that Daniel was the first person to link Newton's discoveries about gravity with Leibniz's calculus, a brilliant step in the development of physics. Without his father's prejudices, Daniel was able to take advantage of the best parts of both Newton's and Leibniz's work.

"Bernoulli, am I correct in thinking that your father isn't very supportive of you in general?" Euler asked.

"No, he never has been," Daniel said sadly. "He was always fondest of my brother Nicolaus, but who could dislike Nicolaus? Sometimes I wonder if my father blames me for Nicolaus' death. I trained as a physician, yet somehow I allowed Nicolaus to die on my watch. I can't help wondering if Nicolaus would have died if he had remained in Bern. Remember, my father didn't choose for him to come to St. Petersburg—only me. According to my father's plans, Nicolaus was supposed to make his career in law at the university in Bern, although he didn't vehemently oppose Nicolaus' plans to come to St. Petersburg. I really don't think he blames me for Nicolaus' death."

"No, I don't see how he can," Euler said. "And if Nicolaus hadn't come with you to Russia, the two of you wouldn't have had the pleasure of all the work you did together here."

"You're right," Daniel said. "We did have eight wonderful months working together here."

"How could your father blame you anyway?" Euler asked. "There is a limit to what a medical doctor can do, and I'm sure you did everything possible. Surely you don't blame yourself for his death, do you?"

"No, honestly, I don't," Daniel said. "For several months Nicolaus suffered from stomach pains, and we consulted with all the medical men in town to try to get help for him. I even wrote to my friend Michelotti in Venice to see what he could suggest. Nothing that any of us could think of actually helped."

"Well, then, it sounds as if you really did all that you could have done," Euler said.

"Yes, I think I did," Daniel said. "Now my father seems to be wondering if I should return to Basel immediately, since it is obvious that the climate here is not healthy."

"In my brief medical studies before I left Basel—and I don't claim to be an expert on medicine by any means—the general wisdom was that for that kind of stomach pain the environment was irrelevant. It would probably have been just as likely to happen in Basel as here. Do you suppose Nicolaus had an open sore in his stomach—an ulcer?" Euler wondered.

"That was my best guess," Daniel said.

"And so the only possible treatment could be a change in diet," Euler said.

"We tried that," Daniel said. "In his last month his diet was as bland as we could manage. He didn't like it, but anything that tasted better gave him agonizing pain. He was miserable."

"I was so sad when I learned that Nicolaus had died," Euler said. "I liked him very much."

"Yes, me too," Daniel said. "He and I were really best friends. I miss him terribly."

"I'm sorry," Euler said. "It must be hard to lose your brother." Then Euler tried to brighten the discussion. "How about your younger brother? Are you close to him?"

"Johann?" Daniel asked. "Not really. I think he's pretty capable although he has always been kind of sickly. He and I are fond of each other, as brothers tend to be, although our friendship is nothing like my friendship with Nicolaus. I was in Italy for several years, at a time when he and I might have been able to form a closer bond. Right now, as far as I can tell, Johann seems to be enjoying studying mathematics and law."

"I've always liked young Johann," Euler said.

"Yes, he's a good boy," Daniel said. "Anyway, here I am in St. Petersburg, and I should admit that I am able to work effectively here. I am particularly looking forward to working with you."

"I can hardly wait to get started," Euler said.

Daniel continued his work on the flow of liquids, discovering what is now called the Bernoulli Principle or the Bernoulli Law, which states that in any moving fluid, as the speed of the fluid increases, the pressure decreases. It is an inverse relation. This is in fact the primary reason that an airplane wing gives lift to the plane. The wing is constructed with a convexly curved upper side and a generally flat underside. The air (which is technically a fluid) going over the top must travel further than the air going under the wing, so it must move faster in order to end up at the back of the wing at the same time. Because that air on top of the wing is moving faster, the pressure on the top of the wing is less, and that means that the airplane is pushed upwards, giving it what we call lift.

Bernoulli's Law also explains why the atmospheric pressure inside a hurricane is so low. The air in a hurricane is flowing at more than 100 miles per hour, and a fluid moving at that speed automatically has a significantly lower pressure than would be found in

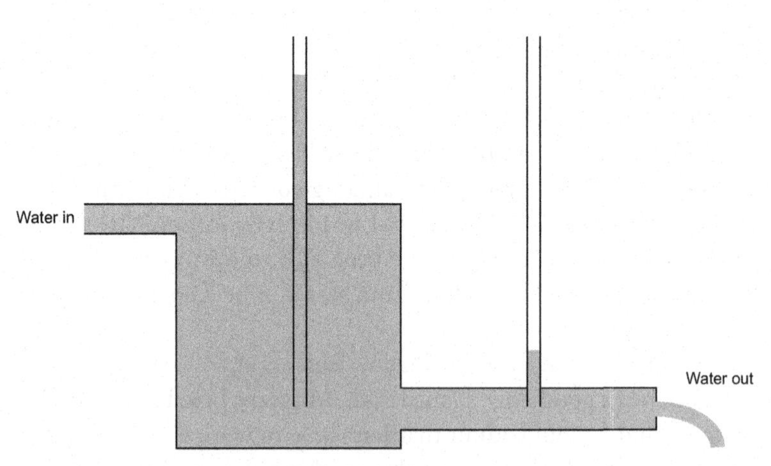

The height of the water in the vertical tubes measures the pressure in the wider tank and the narrower tank, respectively.

a gentle breeze. That explains why roofs of houses are often simply lifted off their houses during a hurricane.

In order to demonstrate his law, Daniel worked seriously with his apparatus. Sealing the seams with beeswax, he carefully inserted narrow glass tubes into the big copper vessel and also into the narrower channel leading out of the big vessel, where he and Euler had noted that the speed of the water was greater. Then he watched the height to which the water rose in the narrow tubes, both in the big part of the vessel and in the narrower channel. Euler, who was a more abstract mathematician, was able to help Daniel prove mathematically what he had observed physically.

Although Daniel completed the manuscript for his book *Hydrodynamica* [*Hydrodynamics* or *Fluid Dynamics*] while he was still in St. Petersburg and left a copy of it with a printer there, he did not submit the book for publication until 1738. By that time, back in Basel, he finally completed the thirteenth chapter. As he did with all his writing, he identified himself humbly as Daniel Bernoulli, son of Johann. The subject of the movement of fluids ever since uses Daniel's name for it: Hydrodynamics.

Euler had a genuine interest in Daniel's hydrodynamics, but he chose to defer to his friend in that field while Daniel was making his important discoveries. Given the quantity of work that Euler produced, he could afford to be generous. In later years when Daniel Bernoulli had gone on to other topics, Euler made significant contributions to Daniel's field of hydrodynamics.

Daniel's study of hydrodynamics would be a happy part of the Bernouillis' story were it not for another unpleasant incident caused by Daniel's father Johann. While it is true that the father and son had done some preliminary explorations of the field many years before, the discoveries of hydrodynamics belong solely to Daniel. In fact, most of Daniel's investigations in the field before his work in St. Petersburg occurred during his time in Venice. However, after Daniel returned to Basel, his father, who knew what Daniel had been doing, secretly wrote his own manuscript on the same topic, calling

DANIELIS BERNOULLI Joh. Fil.
Med. Prof. Basil.
ACAD. SCIENT. IMPER. PETROPOLITANÆ, PRIUS MATHESEOS SUBLIMIORIS PROF. ORD. NUNC MEMBRI ET PROF. HONOR.
HYDRODYNAMICA,
SIVE
DE VIRIBUS ET MOTIBUS FLUIDORUM COMMENTARII.
OPUS ACADEMICUM
AB AUCTORE, DUM PETROPOLI AGERET, CONGESTUM.

ARGENTORATI,
Sumptibus JOHANNIS REINHOLDI DULSECKERI,
Anno M D CC XXXVIII.
Typis Joh. Henr. Deckeri, Typographi Basiliensis.

Title page of Daniel Bernoulli's *Hydrodynamica*.

it *Hydraulica* [*Hydraulics*]. It appears to have been copied in great part from his son's work, and to complete the shameless incident, Johann even falsified the date of publication so that it would appear to have come out before Daniel's book. Daniel was understandably furious at his father's perfidy.

Unfortunately, Johann Bernoulli had also enlisted his former student Euler's help on his dishonest project. He sent some of his work on what he called hydraulics and arranged for Euler to comment on his work. Euler, who couldn't criticize the content and had no idea what was going on between father and son, was quoted as endorsing Johann's work *Hydraulica*. Afterwards Daniel felt that his friend Euler had betrayed him, although Euler, too, was an innocent victim. Johann Bernoulli was a jealous and spiteful man.

27
The St. Petersburg Paradox

"Euler, are you familiar with Montmort's (Rèmond de Montmort, 1678–1719) book *Essay d'analyse sur les jeux de hasard* [*Essay on the Analysis of Games of Chance*]?" Daniel asked his friend one day in St. Petersburg. "Apparently my cousin Nicolaus and Montmort had been in frequent correspondence, and Nicolaus asked him a question on probability that Montmort included in his book."

"Your cousin Nicolaus?" Euler said. "I know Montmort's book, but I have to admit that I have a hard time keeping all your relatives straight. I knew your brother Nicolaus, but this is another Nicolaus?"

"Yes," Daniel admitted, "my family does get pretty convoluted. This Nicolaus is the son of my uncle Nicolaus, who, as you probably know, is an artist in Basel. His son Nicolaus is a mathematician who studied with my uncle Jacob. He originally was trained in the law, like so many of us Bernoullis, and he has worked in logic and law as well as mathematics ever since. He's about 13 years older than I am. He apparently suggested this problem to Montmort."

"Oh, that was a problem about a coin toss, wasn't it?" Euler asked.

"That's right," Daniel explained. "You never forget anything, do you, Euler?"

"Oh, yes I do," Euler assured his friend.

"I've never found you forgetting," Daniel protested. "Anyway, in the game, Peter should toss a coin and continue to do so until it finally lands heads up. Paul waits for it to happen."

"When I read it, I realized that it could go on forever!" Euler observed.

"Precisely!" Daniel agreed. "The rules say that if the coin lands heads up the first time, Peter will pay Paul one ducat. If he doesn't get heads until the second throw, he will pay Paul two ducats. If he doesn't get heads until the third time, he will pay Paul four ducats. Each time he fails to get heads, the premium is doubled. It's really a variation on a problem that my uncle Jacob wrote about when he was first studying probability."

"Interesting," Euler said. "So, Montmort's question had to do with how much Paul should pay for the privilege of playing, didn't it?"

"That's right," Daniel said.

"I think I'd be willing to pay quite a bit if I were into playing such games," Euler said.

"Yes, but Nicolaus doesn't answer the question, and neither does Montmort in his book," Daniel said. "I would think it would be pretty straightforward to calculate the probability of success, wouldn't you?"

"Yes, I think you're right," Euler said.

"But Montmort also proposes inserting a third person into the game, and then considering the general situation in this game of chance," Daniel said. "The third person would serve as the judge."

"I remember that," Euler said. "So in fact you are not talking about a specific game with Peter and Paul?"

"No," Daniel said. "The question is, in general how much should a person be willing to pay to play this kind of game?"

"You could figure it out," Euler said, and noticed that Daniel was already hard at work.

The next day Daniel greeted Euler with the news: "I would be willing to pay any amount at all to play that game—even a million ducats!"

"I thought the chances looked pretty good," Euler said. "Let me see your work."

"Here it is," Daniel said. "It wasn't very difficult."

"Yup," Euler said. "Yes, I have to agree. I'll pay you 100 Rubles to play right now, Daniel. Are you game?"

"Of course not!" Daniel said. "I don't actually play games of chance, and I doubt that you do either. Probability is fine, but still it comes with no guarantee of success on any one individual trial."

"You're not dumb," Euler observed. "The only way to avoid losing at games of chance is not to play them. Why don't you write it up? Don't you think the Academy press would be willing to publish it?"

"I think so," Daniel said. "Our new press is really eager to attract the notice of scientists throughout the world. As you've noticed, they encourage us to submit as many papers as we can."

"We could call your problem the St. Petersburg Paradox," Euler suggested.

"Yes, it is paradoxical," Daniel said. "Hmmm. What do I smell? Are you hungry? It smells as if dinner is just about ready."

"It does smell good. You know, Daniel," Euler said, "it seems to me that we have a pretty good situation here. They pay us to do the work we would do anyway, and we have the privilege of living comfortably and eating well!"

"Yes, in many respects it is ideal," Daniel said. "However, just wait until winter sets in. There is never enough firewood to make it cozy enough for me, and the candles can't begin to compare with the bright sun that shines in Basel."

"You really do want to return to Basel, don't you?" Euler asked.

"I certainly do," Daniel said fervently. "However, getting back to the St. Petersburg Paradox, I'm thinking of inserting something else in my essay. Nicolaus may be surprised, but we might as well consider the psychology of the game as well as its simple probability. I doubt that Nicolaus would have any interest in that aspect of the problem, but it has made me think about people and their reactions to money."

"That they like it?" Euler asked.

"Not just that," Daniel said. "I was thinking about how they react as they get more and more money."

"What do you mean?" Euler asked.

"It seems to me that as people accumulate more and more money, they react differently to each new addition to their pile of money," Daniel said. "I think each additional Ruble is psychologically worth less to them than the Ruble that came before, so as they get richer, the reward is less and less satisfying, so it takes more and more money to please them."

"You are probably right about that," Euler said. "Have you gotten as far as a specific formula for the psychological value of each additional Ruble?"

"I'm getting there. I know the value to the individual of each new Ruble is inversely proportional to the amount of wealth he already has," Daniel said, "although I don't have the precise relation yet. Maybe I'll call the value he places on that additional money its moral value."

"I like it, Bernoulli," Euler said. "So you are using the word *moral* in the same way that your uncle did when he discussed moral certainty."

"I think so. I'm still working on it," Daniel continued. "That moral value must also relate to the way in which the person earns the money and what his expectations are as he works toward it. It's a pretty complex issue."

"I think you could have some fun with it," Euler observed. "I'll be interested to see what you come up with."

"Me too," Daniel said humbly. "I would like to be able to quantify it. I'm really just at the beginning of my thinking about it."

28
Euler's Early Work in St. Petersburg

Euler gave his first lecture at the academy when he was 21 years old in 1728, only two months after he had arrived in St. Petersburg. At the time he was an untried scholar facing an impressive, erudite audience. The problem that he was addressing had been posed by his friend Nicolaus Bernoulli—Daniel's older brother—before his untimely death. Euler intended it as a tribute to the brilliant Nicolaus, who would certainly have addressed the problem himself if he had lived long enough.

The problem was to find a specific curve—a reciprocal trajectory to a given curve. Euler's solution involved a perfectly executed and complicated geometrical drawing, and his calculations resulted in clear formulas. According to Nicolaus' plan, the new curve should have intersected the original curve at right angles, although Euler chose to generalize that requirement to say that the two curves should instead intersect at some unspecified constant angle—not necessarily at 90°. In his lecture, admitting that there are many curves that would satisfy the requirements, Euler decided to limit himself to only the simplest ones. Any mathematician desiring a more complete solution could derive the more complex curves using Euler's work, which established the basic principles that would be needed.

That day in 1728, Euler's audience was struck by several things as the young scholar spoke. He gave his explanations in clear and concise Latin (the working language of the Academy), and he

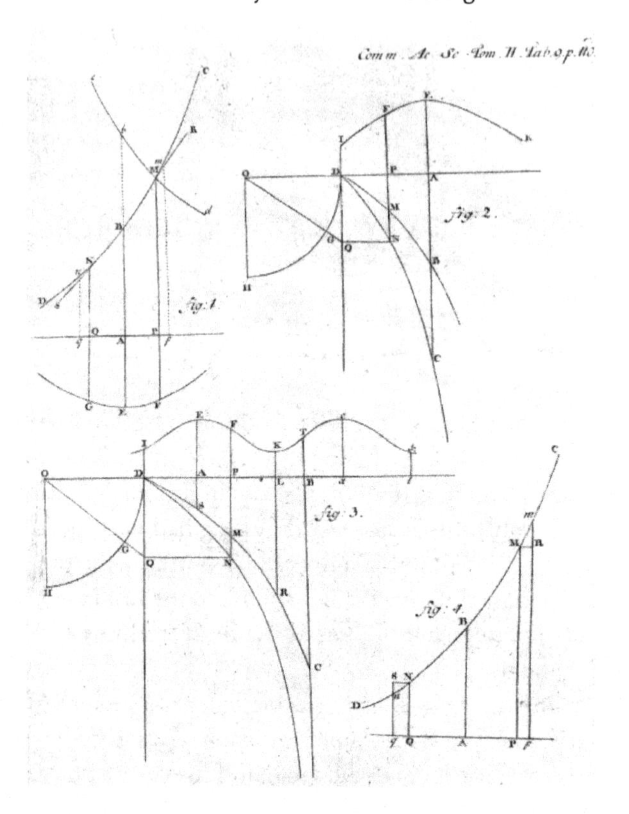

DE *TRAIECT. RECIPROCIS.* 91

motae, mutua interſectione vbiq; angulum eundem conſti-
tuunt, Problema in Act. Lipſ. Suppl. T. VII. a beate hic de-
functo Nicolao Bernoulli propoſitum Ita autem cum hoc
problemate res ſe habet, vt infinitae, tam algebraicae,
quam tranſcendentes curuae ſatisfaciant. Quapropter ad
plenam eius et perfectam ſolutionem requiritur, vt exhi-
beatur methodus, qua curuae ſatisfacientes omnes inueni-
ri queant, ſimpliciſſimae autem tam algebraicae quam
tranſcendentes re ipſa eruantur.

Euler’s reciprocal trajectory curve in St. Petersburg Academy journal.

addressed the question directly. Then he presented some interesting related information, carefully defining what he called odd functions and even functions, which students of analysis have been studying ever since. In his talk, Euler also demonstrated a charming but quiet sense of humor. He allowed his audience to chuckle with him at the same time that he presented very difficult and new mathematics.

After Euler's presentation, his friend Daniel hurried to the lectern to congratulate him. "Euler, that was wonderful," Daniel said, "and I do appreciate the fact that you chose to honor my brother with your work."

"Well, Bernoulli," Euler replied, "it is a fascinating problem and I'm sure your brother would have presented his own solution to it, if only he had had the time."

"Yes," Daniel said, "Nicolaus was a brilliant mathematician."

"I imagine he would have approached this problem in a totally different way than I did," Euler said, "and I must admit that I would love to know how he would have done it."

"We'll never know," Daniel said, "but I think Nicolaus would have approved of your solution, and I believe the members of the academy were impressed by your talk. Since you are young, they may have doubted that you would perform well enough, but I think you passed that test easily. I heard some murmurs of approval as you made your points in your talk. I think they liked it. I know I did!"

"Thank you for your support," Euler said. "I have to confess that I was a bit nervous."

Daniel stepped back to allow the other members of the academy to congratulate Euler on his presentation. Jacob Hermann was particularly eager to talk with Euler and was quick to encourage him in his work. As an accomplished mathematician, he understood Euler's work better than most of the audience, and he found it brilliant.

When Euler wrote his first published essay on this same problem for the academy's journal *Comentarii academiae scientiarum imperialis Petropolitanae* [*Academic Commentaries of the Imperial Academy of Sciences at St. Petersburg*] that same year, his drawings once again were beautifully executed, and his readers, like his listeners, were able to follow his reasoning perfectly. Although St. Petersburg was far from the academic center of Europe, quite naturally it chose to publish its journal in Latin, a language that all scholars could read. The St. Petersburg Academy was determined to be recognized as an enlightened center of learning.

This journal would continue to publish Euler's works regularly for 100 years. In fact, the only issue of that journal between 1729 and 1830 that did not include an essay by Euler was in the year 1793, and that was only because the entire issue was devoted to the memory of Catherine the Great, leaving no room for any scientific essays. The journal was still publishing Euler's work on a regular basis 47 years after his death. Euler was certainly the most productive scientist and mathematician of his time—possibly of all time. Euler was responsible more than any other scientist for the fine reputation of the St. Petersburg journal.

Euler's second article in the academy journal dealt with a process for reducing a complicated equation to a simpler form. In this case, he was concerned with second-degree differential equations, explaining how to reduce them to first-degree equations, thus simplifying the solution process. It was in this article that he presented a novel method of solving a first-degree differential equation through the use of what he called an "integrating factor." It was a breakthrough that calculus students have been using ever since. Euler was a scientist who was clearly working at the cutting edge of his many fields.

In 1727, Euler had begun working on the Königsberg Bridges Problem. The problem featured the town of Königsberg (now called Kaliningrad) in Russia—located on a large island in the middle of the river Pregel at the point where two streams come together to form the river. The legend describes seven bridges that connected the four areas of the town. The residents of the town, who enjoyed taking a leisurely stroll through the town on a Sunday afternoon, wanted to know if it was possible to take an uninterrupted walk that would allow them to cross all seven bridges without ever crossing the same bridge twice.

"Bernoulli," Euler said to his friend one evening after supper, "do you remember the Königsberg Bridges problem?"

"I read something about it once," Daniel said. "It didn't look to me as if the good residents of the town could take their desired walk, but I'm not prepared to defend that view. I take it that you have been contemplating it lately."

"Well, I was thinking about it last night before I went to sleep," Euler said, "and I think I've got a solution. The trip around the town is impossible without crossing at least one bridge twice—as you suspected—but I believe I've got a good solution to a more general problem instead of just a solution to that one problem."

"Okay," Daniel said. "How do you propose to do it?"

"Here, take a look at my drawing," Euler began. "You can see that there are four areas that are connected by those seven bridges. The critical thing is how many bridges serve each area of land."

'That's clear to me," Daniel said.

"Good. Now, when I count the number of bridges serving area *C*, there are three of them–bridges *c*, *d*, and *g*."

"That's true," Daniel said. "Area *A* is served by five bridges–*a, b, c, d*, and *e*–if I am not mistaken."

"Correct," Euler said. "And area *D* is served by three bridges–*e, f*, and *g*–and area *B* is served by three bridges–*a*, *b*, and f. Now, what is significant is that each of those areas is served by an odd number of bridges."

"Yes, they are all odd numbers," Daniel agreed.

"What I have realized is that if there are more than two areas that are served by an odd number of bridges, the trip is not possible with the limitation that you may cross each bridge only once," Euler explained.

"So, the reason that the trip is not possible is that there are four areas served by an odd number of bridges," Daniel said. "What hap-

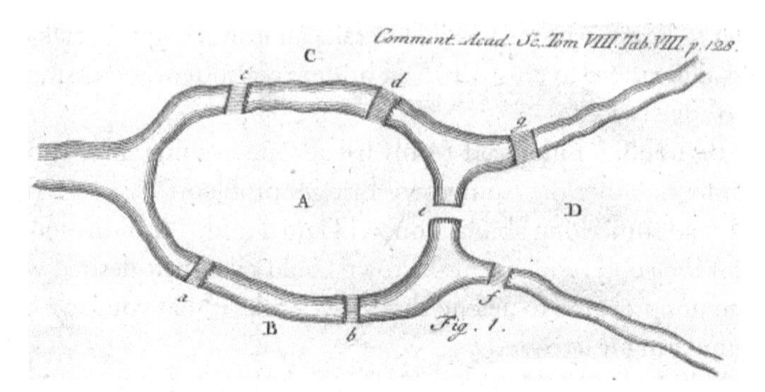

The Bridges of Königsberg problem: A problem in graph theory. Euler's sketch from the *Commentaries of the Academy of Sciences at St. Petersberg.* If and only if no more than two areas are served by an odd number of bridges, it is possible to walk around the town without ever crossing the same bridge twice. However, since each of the areas A, B, C, and D is served by an odd number of bridges, the trip is not possible. If we were to construct another bridge at the left side of the drawing, then areas B and C would be served by four bridges, and the trip would be possible if one starts at A and ends at D or vice versa.

pens if exactly two areas are served by an odd number of bridges? Is the trip possible then?"

"Yes," Euler said, "but only if the walk begins on one of the areas served by an odd number of bridges and ends up on the other area served by an odd number of bridges."

"Are you sure it works then?" Daniel asked.

"Oh, yes," Euler said. "If we build an eighth bridge crossing the river at the left of my drawing, then areas *C* and *B* are served by four bridges. Since *A* and *D* will then be the only areas served by an odd number of bridges, if the trip begins at *A* and then ends at *D* (or vice versa), then the trip is possible."

"Let me try it," Daniel said, and Euler watched while he made the trip around town. "Yes, Euler, you are right," Daniel said, "but it doesn't look like any mathematics that I have ever seen before, Euler," Daniel said. "I think I like it."

In 1735 at the age of 28, Euler presented his solution of the Königsberg Bridges problem to the St. Petersburg Academy. Euler explained how to find a solution to any geometry problem involving a specified number of areas and a specified number of bridges connecting them, calling it *geometria situs* [the geometry of position]. This work is often considered the starting point of graph theory, a part of mathematics that has applications in many different fields—from economics to internet search engines. Euler published his paper on the Königsberg Bridges problem a year later in 1736, demonstrating conclusively that the Sunday stroll in the town with the seven current bridges was impossible. Königsberg's eighth bridge was eventually built in 1875, perhaps gratifying that mathematical need of the residents.

In 1736 at the age of 29, Euler published his first book, the two-volume *Mechanica.* In 1739 at the age of 32, Euler published the book, *Tentamen novae theoriae musicae* [*Essay on a New Theory of Music*]. In this book, Euler used logarithms to explore musical scales as he compared the frequencies of the vibrations of a string. He and his friend Daniel Bernoulli often discussed the physics of music at this time, exploring the harmonics of a tone when the string vibrates as it is played on an instrument.

When they were first working together on music, one of Bernoulli's first discoveries was that when a string vibrates to produce a given tone, in fact it has many different frequencies all at the same time. Those multiple frequencies give the sound a richer flavor, allowing sympathetic vibrations to occur in response to a given note. The music in early video games did not have those secondary frequencies, thus explaining their relatively flat sound.

"Euler, listen to this," Daniel said to his friend one afternoon. "I'm holding down the note C above middle C on the clavier, and now I'm going to strike the middle C. Listen carefully."

"Remarkable, Bernoulli!" Euler said. "I hear the higher C even though you didn't strike it."

Euler and Daniel Bernoulli's workplace must have been cacophonous sometimes as they conducted their music experiments, although at other times the music must have been quite enchanting. Euler was an accomplished musician who took great pleasure in playing the clavier throughout his life.

Euler's book on music suggests an interesting parallel to Johann Sebastian Bach's great work *Das wohltemperierte Clavier* [*The Well-Tempered Clavicord*], which was published in 1722, 17 years before Euler's book and explored the full range of keys on the clavier, presenting a prelude and a fugue in each of the 24 major and minor keys. Euler and Bach between them revolutionized music in the eighteenth century. Although Euler's work has been criticized as having too much music for mathematicians, and too much mathematics for musicians, it was an important step in the analysis of music through physics. It is interesting that in their own time, Euler was far better known than Bach—any reasonably well educated person had at least heard of Euler while virtually no one outside of his immediate geographic area had ever heard of Bach.

When we look at Euler's mathematical output, we may be tempted to say that its outward appearance really doesn't look exotic at all. His notation is basically the same as ours. While that is true, it is true for an interesting reason: His notation looks like ours because we use his notation. For example the symbols π, e, Σ, γ (gamma), and the practice of naming the sides of triangle *ABC* using the lower case letters *a*, *b*, and *c* to stand for the sides opposite the angles *A*, *B*, and *C*, respectively, either originated from Euler's pen or became standardized because he used them. It is because of his pioneering work that we still use them.

Suggesting that Euler really wasn't original is similar to saying that Shakespeare's writing is trite—the reason Shakespeare uses expressions that we find hackneyed is because when we use them we are actually quoting Shakespeare. It is the same with Euler, and it is no exaggeration to compare him with Shakespeare. As Pierre-Simon Laplace (1749–1827), the great French mathematician, wrote many years later, "Read Euler; read Euler; he is the master of us all."

In addition to his scholarly work, Euler, like all the professors at the academy, had other responsibilities. He taught courses in physics, logic, and mathematics at both the secondary school and the university levels. He also was involved in examining members of the cadet corps, and he assessed the competence of graduates in geodesy—map making and map reading. He studied carefully the precise weights of the academy's system of weights for the government's measuring. He consulted on the construction of a saw mill and methods to improve the pumps for the city fire department. He designed a satisfactory steam engine that was used to provide water pressure for the pumps of the fire brigade. On top of all this mundane work, Euler continued to produce new works in mathematics and physics at an astonishing rate.

29
Daniel Returns to Basel, and Leonhard Euler Becomes Professor of Mathematics at St. Petersburg

In 1733 at the age of 33, Daniel Bernoulli was finally offered a professorship in botany at his home university, the University of Basel, after several earlier failed attempts. Certainly, botany was not the subject that he would have chosen, but it was good enough. Most important, it was in Basel. He correctly expected to be able to trade it for a better field later, although his uncle Jacob Bernoulli probably would not have approved of his opportunism. Although Daniel was reluctant to leave behind in St. Petersburg his friend Leonhard Euler, with whom he had worked so effectively for several years, he hated the harsh climate in Russia more than ever, and he yearned for his native Basel. How could a scientist who took such pleasure in the movement of fluids bear to be stranded in that frozen wasteland for so many months of the year?

When he had first traveled to St. Petersburg, Daniel had done it in the company of his older brother Nicolaus, with whom he was always very close. When his brother had suddenly died, Daniel had been devastated, but Euler arrived the following year to work congenially with him and to provide the intellectual stimulus that he had earlier enjoyed with his older brother. Although Daniel's St. Petersburg years working with Euler turned out to be his most productive, he had every reason to expect that his productivity would only increase once he returned home to his native Basel.

When he got the call to return to Basel, Daniel joyfully invited his brother Johann, who was then 22 years old (ten years younger than Daniel), to come and work with him and Euler for a time in St. Petersburg, after which the two brothers could then make the journey back to Basel together. Although young Johann, following his father's instructions, had studied law and obtained his doctorate in jurisprudence during Daniel's absence, in fact, Johann actually pursued a career in mathematics like so many Bernoullis. For unknown reasons, Johann's father didn't perceive his son Johann as the threat that his son Daniel seemed to pose to the distinguished mathematical patriarch.

"Johann," his father Johann Bernoulli said to his younger son when he heard about young Johann's proposed trip, "I really don't think a trip to St. Petersburg is advisable for you. It will be a long and arduous journey. It is a very long way from Basel."

"Oh, Father," Johann protested, "I'm sure I can survive the trip, and it would be such fun to see where Daniel and Euler have been working."

"But Son," his father said, "You are not as strong as those two young men. The journey at sea alone will be strenuous."

"I know that, Father," Johann said, "but I'm sure I'm strong enough to make the trip."

"Your mother and I were talking about it last night, and we are both concerned," his father continued.

"You allowed my brothers to travel to St. Petersburg when they were my age, and I am determined to go," Johann said.

Johann's journey to St. Petersburg took him through the major cities of Prussia, before the final leg of the trip took him across the Baltic Sea to St. Petersburg. Remaining with Daniel and Euler in what Daniel enjoyed referring to cynically as the "Venice of the North" for a little more than a year, Johann enjoyed working together with his brother Daniel and his friend Euler, pursuing many questions of mathematics, astronomy, and physics with them. Johann saw for himself what a congenial place the St. Petersburg Academy was at the time—with its collection of brilliant scientists and its modern facilities.

Daniel was pleased to find that his younger brother had matured as a mathematician, and he looked forward to working with him when they got home to Basel. Johann was certainly not the intellectual equal of either Daniel, or of their brother Nicolaus, or of their friend Euler, but he was competent and congenial.

On a personal level, young Johann was able to bring Daniel and Euler up to date on all that had been happening in Basel in the years since they had left. One of the more interesting stories that he told them concerned Pierre-Louis Moreau de Maupertuis (1698–1759), who had spent several months in the Bernoulli household studying Leibniz's calculus with their father at the same time that young Johann was learning it. As a nobleman, Maupertuis assumed that he could simply present himself in Basel, the logical location to study Leibniz's calculus. He expected to receive the instruction he desired—and it turned out that he was right. Since the elder Johann Bernoulli still welcomed the additional income he earned from visiting students, and Maupertuis was a very capable student who also had the funds to pay for his instruction, it was a satisfactory arrangement for both of them.

During those months, young Johann and Maupertuis had become good friends, and Daniel and Euler were eager to hear about the young nobleman's plans. Maupertuis would soon use his skills in the calculus on his 1736 journey to Lapland in the arctic, where he was determined to settle the question of the shape of the earth for his king Louis XV—is the earth a flattened sphere like a tomato or an elongated sphere like a lemon? After the arduous trip, his calculations convinced him that it resembles a tomato (as he had expected), and his conclusions were eventually accepted.

Daniel and Johann set out by ship from St. Petersburg in the summer of 1733, and Daniel and Euler corresponded regularly from then on. In one letter written from Paris on September 22, 1733, Daniel discussed some discoveries he made during the voyage across the Baltic on a large sailing ship. This was a perfect opportunity for him to witness the use of wind in propelling a ship—in other words a perfect laboratory for exploring fluid dynamics—and he was surprised at the results. Since the friends were equally comfortable in German, French, and Latin, the choice of language was arbitrary as can be seen—he started in German, but then switched to Latin several times and then even threw in a little French before returning to his native German. The words that are written in regular font are in German, *those in italics are in French*, and THOSE IN SMALL CAPITALS ARE IN LATIN. He wrote,

> Ich habe auch gesehen die wahre Ursach, warum das Schiff CAETERIS PARIBUS geschwinder geht mit halbem Winde, als mit vollem Winde. Die Ursach ist gar nicht, wie man bishero geglaubt, dass man alle Segel mit halbem Winde employiren könne; denn die OBLIQUITAS VELORUM derogirt mehr als man A NUMERO VELORUM gewinnt, welches gewiss ist.Die wahre Ursach ist, dass mit einem *vent en poupe en faisant force de voile,* das Schiff schier DIMIDIAM VELOCITATEM VENTI, oder auf das wenigste TERTIAM EJUS PARTEM erlangt. Weil nun die RATIO VELOCITATUM notabel ist, so ist VELOCITAS RESPECTIVA VENTI bei einem halben Winde viel grösser als bei vollem Winde, und kann also in dem ersten Falle das Schiff geschwinder getrieben werden, als in dem andern. Aber die Zeit lässt mir nicht zu, von dergleichen Materien weitläuffiger zu seyn.

In English it means,

> I have also seen that the real reason why a ship, ALL OTHER THINGS BEING EQUAL, goes faster when the boat is at an

> oblique angle to the wind rather than with the wind straight behind. The reason is not, as was previously believed, because the half wind enables one to use all the sails; in fact, WITH THE OBLIQUE SAILS, one loses more than one gains with multiple sails, which is a fact. The real reason is that with a *wind from behind giving power to the sails*, the ship barely ACHIEVES HALF THE WIND SPEED, or even as little as A THIRD OF IT. Because THE RATIO OF THE SPEEDS—wind speed to boat speed—is important, THE RELATIVE WIND SPEED with a half wind is much greater than with a full wind, and a ship can thus be propelled faster in the first instance than in the second. But time does not permit me to be more expansive in such things.

As they had worked together in St. Petersburg, Euler and Daniel had undoubtedly switched languages in a similar fashion, and they would hardly have noticed the change either in conversation or in correspondence.

Daniel also mentioned in his letter that he had word from Basel that the chair of eloquence at Basel was open. Although he admitted that he personally had no interest in that field, he then suggested that perhaps his brother Johann should try for that position, which he apparently did successfully. Before closing the letter, Daniel told Euler that he had learned that their friend and colleague in St. Petersburg, the Swiss Jacob Hermann, a distant relative of Euler, had recently died in Basel. That was sad but important news for Euler.

Daniel and Johann's trip back to Basel, first by sailing ship and then by horse-drawn carriage, lasted two months. Daniel had always been aware of how remote St. Petersburg was from what he considered the civilized world, but the length of the trip emphasized that fact once again, although they did repeatedly stop along the way to meet and talk with mathematicians throughout Germany and France.

As Daniel had expected, on that long journey he also had to constantly watch out for his brother to be sure that he was not over-

Daniel Bernoulli.

tired. When they finally arrived back in Basel, Daniel was relieved. He had succeeded in delivering his younger brother in reasonably good health back to his parents. Perhaps more importantly, Daniel was finally home. He had dreamed of this homecoming for almost eight years.

Before leaving St. Petersburg, Daniel had submitted an essay for the Paris Prize competition of that year, concerning the effect on planetary rotation that would result in an increase in the tilt of a planet's orbit toward the equator of the sun. Soon after he arrived home, Daniel found a letter from Paris, addressed to both him and his father.

"Hmmm," Daniel said to himself. "That's odd! I wonder why it is addressed to both of us. Do you suppose we could have both won? Wouldn't that be magnificent!"

He opened the letter and found that indeed they had both won, and he waited impatiently for his father to return home from work that day. As soon as Daniel saw him, he proudly presented the envelope to him. When his father seized it, 33-year-old Daniel saw with horror that his father's expression did not mirror his own. His father, at the age of 66, was livid.

"How could you have done this to me?" his father demanded.

"But we both won," Daniel said. "Don't you think that is wonderful?"

"No!" his father said. "It is wrong for you to be treated as my equal! You are my child, who has never been able to do anything without my help. Get out!"

"What?" Daniel asked incredulously.

"I said, get out! This is not your home anymore," Johann said. "I wish you had stayed in Russia."

Daniel was devastated, and the rift would never heal.

"Johann," his wife Dorothea (Daniel's mother) asked later that evening, "what is the matter? Where is Daniel? I thought he would join us for supper."

"The Paris Prize—that's what's the matter!" Johann roared.

"Did you win the prize, Johann?" Dorothea eagerly asked.

"Yes, but I have to share it with Daniel," Johann fumed.

"So that means that you have won it for a third time," Dorothea said, "only this time Daniel has won too! I'm so proud! Everyone seems to respect you as the greatest mathematician in all of Europe, and now Daniel is developing a fine reputation as well. That makes me the happiest woman in Basel! How can you be angry about that?"

"Because I have to share it with that upstart," Johann said. "Our son Nicolaus would have been my peer. Daniel is not."

"But Daniel is a brilliant young scientist," Dorothea reasoned.

"He is not my equal," Johann shouted. "This is the greatest insult I have ever suffered."

"Oh, Johann," Dorothea said, "I can't believe you are reacting this way. Daniel is achieving respect as a mathematician in his own right, and I'm so pleased to have him home again. I am so proud."

"He will never live this down," Johann announced. "I will never speak to him again."

In 1744, Daniel's brother, 33-year-old Johann Bernoulli, now professor of eloquence at Basel, was married to Susanna König, also of Basel. She was the daughter of a professor of Greek at the university. Johann and Susanna had eight children, several of whom Johann instructed in mathematics and three of whom—Johann, Daniel and Jacob—became successful mathematicians, though not of the caliber of Daniel and Nicolaus. When his father died, Johann was chosen to occupy his father's chair of mathematics at the university.

As a young man, Johann did some impressive mathematics, winning the Paris Prize either by himself or jointly with his father four times. After his father's death, however, Johann's work was not as impressive as it had been earlier. However, despite being shy and apparently frail, young Johann lived to the age of 80. He apparently did not engage in the academic battles that were typical of his father, although, like his father, he was a prolific correspondent with mathematicians all over Europe.

Johann's son Johann, Daniel's nephew, was recognized as a child prodigy but in fact accomplished little as a mature scientist. His only writing that had lasting importance is found in his travel diaries from 1772 to 1781 as he explored parts of Germany and Poland. Although Frederick the Great invited him to go to Berlin to reorganize the Berlin Academy, his frail health kept him from accomplishing that.

In 1739 at the age of 29, Daniel's brother Johann would travel with Maupertuis to Paris. On the way, the travelers stopped at Cirey, where Emilie, the Marquise du Châtelet, shared her manor and engaged in witty conversation with Voltaire (1695–1778), the erudite wit who entertained and scandalized high society in Europe at the time. Since du Châtelet and Voltaire were enthusiasts for Newton's work (du Châtelet would translate Newton's *Principia* into French a few years later), young Johann's father Johann probably did not approve of this detour to visit with Newtonians on the way to Paris, but Johann and Maupertuis apparently enjoyed their visit.

Many years later, when Maupertuis was old and sick and had long since been applauded for his conclusions about the shape of the earth, he returned to Basel and stayed with his friend Johann Bernoulli (Johann's son and Daniel's younger brother) in his final months. In his poem "The Death of the Hired Man," Robert Frost observed, "home is the place where, when you have to go there, they have to take you in." The Bernoullis apparently provided that refuge for Maupertuis.

30
Daniel Bernoulli: A Famous Scholar

In all, Daniel Bernoulli won the Paris Prize ten times—many more times than his father Johann did—although the comparison is not entirely fair since the prize had been established only in 1721, when Johann was already 54 years old. It is well known that many brilliant achievements in mathematics are the work of scientists who are in their teens or their twenties—mathematics has sometimes been called a young person's sport. Jacob and Johann Bernoulli would probably have won the prize several times for their work on the development of the calculus and probability when they were young men, if the prize had been available then.

Johann's son Daniel had won the Paris Prize for the first time in 1725 as a young man working in Italy before going to St. Petersburg, and then again in 1728, soon after he began working in St. Petersburg. After the 1734 prize, which he "shared" with his father, he won it in 1737 for designing the optimal shape for a ship's anchor, sharing the prize this time with an Italian physicist named Giovanni Poleni. International trade and shipping provided much grist for the scientific mill at that time.

Daniel triumphantly shared the 1740 prize with his friend and colleague Euler and several other scientists for investigations concerning the tides. In 1743, he won the prize for a work on magnetism, and in 1746 he was glad to share the prize with his younger brother Johann for another work on magnetism. In 1747, he won the prize with a plan for his complex pendulum, in his quest for the solution to the longitude problem. In 1753, he won the prize for an essay dealing with the effect on ships of forces related to the wind. In 1757 at the age of 57, he won the prize for the last time with practical proposals for reducing the roll and pitch of a ship at sea. All these projects on navigation were related to his work with fluids—of both air and water—placing him in the forefront of hydrodynamics. His father apparently had only minimal further interest in that field after his failed attempt at competing with his truly gifted son and adversary.

By now, Daniel Bernoulli had earned respect as a great scientist in his own time even if his own father never acknowledged his accomplishments. In later years, Daniel enjoyed telling of a chance encounter a few years after his return to Basel from St. Petersburg. He was traveling by coach and entered into conversation with another passenger during the several hours that they traveled together. The men enjoyed one another's company, and during a lull in the conversation, Daniel realized that he hadn't yet introduced himself.

"I'm so sorry," Daniel said. "I should have introduced myself an hour ago when we started to talk. I am Daniel Bernoulli."

The man stared at him in astonishment for a moment. Finally he regained his poise and said, laughing at the wonderful joke, "Yes, and I am Sir Isaac Newton! Ha! Ha!"

"But Sir," Daniel said in some embarrassment, "I truly am Daniel Bernoulli." Daniel, always a humble man, was both astonished and immensely flattered to be presumed to be on a level with the great Isaac Newton.

"But you are too young!" his companion said. "I expected that Professor Bernoulli must be a grizzled old man. Are you really the author of those brilliant books and the winner of so many prizes from the Paris Academy?"

"I had remarkably good fortune in my early career," Daniel explained. "My brother Nicolaus taught me mathematics when we were young. I had the privilege of working closely with him and then with Leonhard Euler, whom you have undoubtedly heard of, in St. Petersburg, and for the last few years I have been able to continue my work in Basel in the beautiful Helvetian Confederation."

"I commend you, Sir," his companion said. "You have accomplished great things!"

Several years after that incident, Daniel was invited out to an evening meal in the company of several of Basel's leading citizens. Toward the beginning of the meal, the host explained a mathematical problem that was stymieing him—he frankly admitted that he didn't have a clue how to solve it and yet it was the key to his current project. Daniel, along with the other guests, listened carefully but made no comment. The discussion then went on to many other topics of the day, and Daniel participated happily in the talk. However, when it was time for Daniel to go home, he thanked his host for a delightful evening and then slipped a piece of paper with some small calculations into his host's hand.

"What is this?" the host asked.

"That's the solution to your problem," Daniel explained. "I jotted it down for you during a lull in the conversation. I hope you will be able to use it. If you'd like some more information, don't hesitate to ask."

"But I've been struggling with it for a week!" the host protested.

"Yes, but that isn't the kind of problem that you devote most of your time to," Daniel said. "If it were, you could have solved it as easily as I did."

"Thank you so much!" his host exclaimed. "That will save me a great deal of time. I will be sure to give you credit for your wonderful help."

"Don't worry about it," Daniel demurred. "It's not important. Feel free to use it any way you like. Thank you for the delicious meal and such an enjoyable evening."

In 1743 at the age of 43, Daniel was able to change from the chair of botany to the chair of physiology, and then finally in 1750 to the chair of physics, which he held until his death in 1782. His lectures in physics, which were accompanied by fascinating demonstrations—carrying the subject far beyond the lectures his uncle Jacob had delivered so many years earlier—held the complete attention of all his listeners year after year. In his lectures, Daniel often presented his latest research, which he often had not yet completely proved, although over time all his discoveries have been accepted as valid. His students were captivated by all that he shared with them, happy in the knowledge that they were the first people to learn of their amazing professor's brilliant discoveries.

In 1760, Daniel presented a paper using probability to the Paris Academy on the revolutionary concept of inoculating a population against smallpox. Bernoulli calculated that at that time approximately 3/4 of the population of Europe had been infected with smallpox, and that fully one-tenth of all deaths at the time were due to smallpox. He reported that anyone who had survived smallpox had a resulting immunity to the disease, and based on his statistics he recommended a process called variolation, deliberately inducing what was then called "artificial smallpox" in a subject so that the person suffered a milder infection that would later protect him from the serious ailment. Although there was a slight risk—this procedure occasionally resulted in death—at least in part because of Daniel's statistical work, smallpox epidemics ceased to be a problem at least in England by the end of the nineteenth century. In 1764, John Adams, the second president of the United States, chose variolation for himself—it made him sick, but not too sick—and he strongly recommended the process. Later less invasive smallpox inoculations were much less risky, but variolation was the first step in eradicating that plague.

When Johann, Daniel's father and Euler's mentor, died in 1748, he was apparently confident in his position as Basel's finest mathematician. The epitaph on his grave in the Peterskirche identifies him as the "Archimedes of his age." Cantankerous though he was, he was the most important mathematician in Europe after his brother Jacob's death and before his son Daniel's and his student Euler's rise to prominence. That gave Johann 25 years to reign as the Prince of Mathematicians.

Daniel Bernoulli never married, and is buried in the same church, along with his famous father and uncle. He is revered as a great scientist, at least on a par with his father, who would have resented that fact.

31
Leonhard Euler: Admired Professor at St. Petersburg

After Daniel's departure from St. Petersburg in 1733, he was delighted to learn that his friend Euler, who was then 26 years old, had been named professor of mathematics in his place. Although Euler had been sad to see his friend depart, Daniel's unhappiness in St. Petersburg had been no secret, and by this time Euler had made many friends, both within the academy and in the town as well. With his promotion came a significant increase in salary, so that Euler could begin to think of establishing himself personally and professionally.

Euler had occasionally talked with a charming young woman, Katharina Gsell, the daughter of a Swiss painter whose family had lived in St. Petersburg for several years. One evening he bravely walked over to the Gsell's house to talk with Katharina's father.

"Good evening, *Herr* Gsell," Euler began. "I hope you are well."

"Good evening, *Herr* Euler—or I should say Professor Euler," his host replied. "I'm delighted to see you."

"Thank you, Sir," Euler said. "I have come to ask you a question that is very important to me."

"I would be pleased to help you in any way within my power," *Herr* Gsell replied.

"*Herr* Gsell, I would very much like to marry your lovely daughter Katharina, if you could approve of me as a proper husband for *Fräulein* [Miss] Gsell," Euler said.

"What a charming thought!" *Herr* Gsell replied. "I have been watching you as you have been making your way in the academy, and it looks to me as if you are headed for a fine career."

"I hope I will be able to live up to your opinion," Euler said.

"I'm sure you will, *Herr* Professor," Gsell said. "Do I understand that your father is a Reformed minister in Basel?"

"Well, actually just outside of Basel," Euler corrected him. "I grew up in the town of Riehen and then I studied at the Latin school and at the university in the city of Basel."

"I would be delighted to welcome you into our family, although I expect we should discuss this with my daughter before we make it final," Gsell said. "Let me see if she's available now. Just a minute please." When Gsell disappeared into the rest of the house, Euler stood there quietly, amazed at his good fortune. If only Katharina would agree ....

Soon Gsell returned with his daughter, explaining to Euler, "I have told Katharina about our talk and she seems to be as delighted as I am."

"Oh, *Fräulein* Gsell," Euler said, turning to the young woman and taking both her hands in his, "would you consider being my bride? I would be the very best husband that I can possibly be."

"Yes, *Herr* Euler," Katharina replied quietly but firmly, "I would like very much to be your wife." Katharina, who was just one year older than Euler, had admired him from a distance for more than a year.

As far as Euler was concerned, the only possible difficulty in this marriage might be the close friendship between Katharina's father and *Herr* Schumacher, the obnoxious dictator of the academy. However, Euler was a diplomat who refused to let his personal feelings sour his relationship with his future father-in-law, and over time the Eulers even went to the occasional party at the Schumachers' home.

The happy couple was married in January of 1734, and they were soon able to move into their own well-furnished and comfortable house, which was within easy walking distance of the academy. Since stone in St. Petersburg was reserved for the construction of the great academy in Tsarist Russia, the Euler house was made of wood, like all the other private dwellings in town. Katharina was always uneasy in that house, beautiful and comfortable though it was, since it was made of wood and was thus susceptible to fire. In St. Petersburg at that time, however, Katharina knew that a wooden house was the only choice.

When Euler was called to supervise the department of geography at the academy in addition to his work in mathematics, his salary was again increased, this time to an impressive 1,200 Rubles. There was no doubt that the academy recognized Euler's phenomenal talent. He and his wife were now sitting in a comfortable position. Katharina gave birth to their first child in November, 1734, naming him Johann Albrecht Euler, after Johann Albrecht Korff, his first godfather and the president of the St. Petersburg Academy at the time. Johann Albrecht's second godfather was Christian Goldbach, friend of the Bernoulli family and Euler. Both godfathers were honored to be chosen for this role.

Johann Albrecht was soon followed by a second son, Karl Johann. Also at that time, Euler's younger brother Heinrich came to St. Petersburg from Basel to live with the young Euler family so that he could study art there (perhaps with Katharina's father), before he moved on to Paris to study there. Katharina and Leonhard's two sons were followed later by 11 more children, although only five of the Eulers' children survived to adulthood. As young parents, the Eulers were familiar with the grief associated with losing young children to disease.

Following the departures of first Jacob Hermann and then Daniel Bernoulli, Euler was glad to work more closely with Georg Wolfgang Krafft, a physicist who was about his age and was a creative, active scientist at the academy. Krafft, who had grown up in Tübingen in the Duchy of Württemberg in what is now southern Germany, happily spoke German with Euler. Although they shared their native language, Krafft unfortunately did not speak it with Euler's distinctive Swiss accent as Daniel had. Krafft and Euler collaborated on many investigations, each probing the other's discoveries with the intelligent insight of a different perspective. Since the fields of physics and mathematics are closely connected, Euler and Krafft spoke the same language scientifically as well as linguistically. Within the academy, Euler considered Krafft his closest friend.

Euler was a tireless writer, focusing on a huge variety of topics within any given week. He had astounding powers of concentration, often working with children playing around him on the floor or a cat on his lap, and apparently able to interrupt his work for a moment or two, and then to return immediately to the writing at hand. Legend credits him with being able to compose a new scientific essay between the first and second calls to dinner!

He wrote essay after essay, simply laying each completed one on the floor next to his chair before beginning the next one. Because Euler conceived of a whole essay before he began to write it, each essay was complete when he finished the first and only draft. At that point it was ready to be delivered to the printer. Often the essays piled up on the floor beside his chair, and, when the printer's assistant at the Journal of the Academy at St. Petersburg came for a few essays, he simply took the ones that were on top. As a result, the order of publication of Euler's work has little to do with the order that he wrote them in—sometimes the essays on the bottom of the pile would wait for months before being delivered to the printer. There was no way that any printer could keep up with Euler's phenomenal output.

Also at this time, Euler and the Bernoullis' friend Christian Goldbach, a German who had grown up in Königsberg, came to know each other well. Goldbach, who was ten years older than Euler, was never considered a serious scholar. Rather, he was a citizen of the world who spoke French, German, and Latin beautifully and was considered erudite in many fields. Goldbach's first assignment in St. Petersburg had been as a tutor to the boy tsar Peter II. After the young tsar died, Goldbach continued on at the academy, serving as corresponding secretary at the same time as he was becoming more and more influential in the government of the Tsarina Anna. Although Goldbach was no mathematician, by this time he was familiar enough with the subject and sufficiently interested in it that he and Euler had much to talk about. Goldbach always asked excellent questions, and Euler had the genius to answer them. Goldbach's correspondence with Euler fills one large volume of Euler's *Opera Omnia* [*Collected Works*].

Although Euler did not begin at once to correspond regularly with his mentor Johann Bernoulli in Basel after his arrival in St. Petersburg, by the 1730s their correspondence was active. One of the interesting features of Johann's letters to his former student is the evolution of the greetings at the beginning of his letters to Euler.

In his first letter to Euler in 1728, Bernoulli opened in Latin with the simple *Doctissimo atque ingeniosissimo Viro Juveni Leonardo Eulero* [to the highly erudite and most ingenious young man Leonhard Euler]. A letter three years later in 1731 begins with a more florid greeting: *Clarissisme et Doctissime Domine Professor, Amice Carissime* [brightest and most learned professor, dearest friend]. By this time Bernoulli seemed to be prepared to recognize Euler with enthusiasm. He had seen the products of Euler's pen, and he was impressed with his prodigy's performance.

A letter in 1737 begins with the greeting, *Viro Clarissimo ac Mathematico longe acutissimo Leonhardo Eulero* [to the very brilliant and enduringly sharp mathematician, Leonhard Euler]. Johann was now openly proud of his student, whom he apparently never saw as a rival.

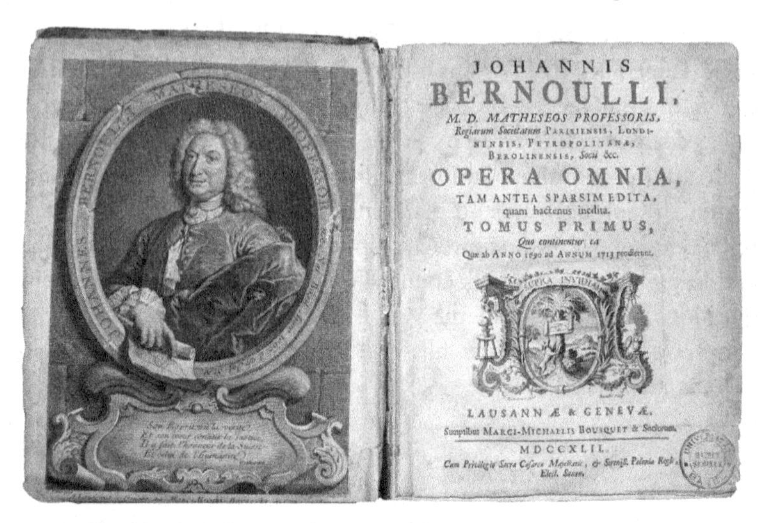
JOHANNIS
BERNOULLI,
M. D. MATHESEOS PROFESSORIS,
Regiarum Societatum PARISIENSIS, LONDINENSIS, PETROPOLITANÆ, BEROLINENSIS, Socii &c.
OPERA OMNIA,
TAM ANTEA SPARSIM EDITA,
quam hactenus inedita.
TOMUS PRIMUS,
Quo continentur ea
Quæ ab ANNO 1690 ad ANNUM 1713 prodierunt.
LAUSANNÆ & GENEVÆ,
Sumptibus MARCI-MICHAELIS BOUSQUET & Sociorum.
MDCCXLII.

Title page of Johann Bernoulli's *Opera Omnia* [Collected Works].

Nine years later, a letter in 1746 begins with *Mathematicorum Principi Leonhardo Eulero* [to the Prince of Mathematicians, Leonhard Euler]. This time he recognizes 39-year-old Euler as the Prince of Mathematicians. With this greeting, he gives Euler the title that he (Bernoulli) had been proud to hold as his own among the mathematicians of his day. In other words, the arrogant and respected king of all mathematicians in the world was bestowing this title on his heir, Leonhard Euler. Johann was certainly demonstrating that he could react to a younger scholar's work with something other than a resounding *No*!

Most of these letters dealt with questions of mathematics that the younger scholar was exploring, but Johann also devoted parts of many letters to news about the priority dispute with the Newtonians in England—Johann's favorite topic for many years.

"Well, Euler, you wouldn't believe how aggressive the English have become," Johann wrote. "They claim that Leibniz had an extensive correspondence with Collins and Oldenburg back in the early 1670s, through which he was supposed to have learned enough about Newton's fluxions to put together the rest. Can you believe it? I am in constant communication with Leibniz through all of this. I am becoming more and more convinced that Newton may actually have stolen the calculus from Leibniz. Leibniz's articles would certainly have been available to Newton if he chose to look at them. It is outrageous!"

Johann seems to have been treating Euler almost as a son. It is a touching tribute, but the contrast to the way he treated his own son Daniel, who was carrying out his own brilliant researches in Basel at the time, is striking. Johann Bernoulli, Daniel's father and Euler's mentor, was a difficult man.

Beginning in 1738, Euler won the Paris Prize three years in a row with essays on a variety of topics, both practical and theoretical. Over the course of his life, Euler submitted a total of 18 essays to the Paris Academy, winning first prize as many as 12 times in addition to once more when his son Johann Albrecht won, certainly with some help from his father. The only person who came close to matching Euler's record was his friend Daniel Bernoulli, who won it ten times. Daniel Bernoulli reveled in his friend's impressive accomplishments at the same time that Euler reveled in Daniel's.

During his years in St. Petersburg before 1741, Euler produced four books: *Mechanics*, *Naval Science*, *Introduction to a New Theory of Music*, and a two-part textbook in arithmetic for high schools, *Introduction to Arithmetic*. The first three books immediately won the respect of scholars throughout Europe, while the arithmetic textbook would continue to be used in the Russian schools for many years.

32
Euler Becomes Blind in His Right Eye

In 1735 at the age of 28, Euler was struck with a serious infection and what was described as a "fiery fever," which almost killed him. Although he survived with no lasting problems, it made him aware that, like all people, he was mortal. He wrote to his friend Daniel Bernoulli, giving him a complete description of his brush with death.

Daniel sent a letter in reply: "Euler, I'm so glad to hear that you have made a good recovery from that infection! A few days after I received your letter, I walked out to Riehen to talk with your parents. I guess I'm not surprised that you hadn't written to them about your illness, and I was glad that I was able to tell them about the infection only when I also had the news that you have recovered. I would have hated to leave them wondering and worrying. You know, your parents and I are greatly relieved that you survived that infection, but in fact the world of mathematics should also express relief. If something happens to you, where would we all be? You alone are essential to the progress of mathematics! I hope something like this never happens again."

Daniel's letter then went on to discuss his current research and writing and to thank Euler for sending him portraits of himself and his wife. "The portraits that you sent me of you and your lovely wife are wonderful. The artist (is it *Herr* Gsell who painted them?) certainly captured your essence, and I think he was similarly successful with your wife Katharina. I'm so pleased that you had her support during your illness. Please give my love to her and your children, Bernoulli."

Three years later at the age of 31, Euler was again struck with a serious infection, which this time led to the loss of vision in his right eye. The portrait of himself that Euler had sent to Daniel (and which has since been lost) was the only formal oil portrait made of him when both his eyes were functioning. Once again, Euler wrote about this illness to his friend Daniel, although apparently not to his parents, who once again learned about it only from Bernoulli.

Daniel wrote back to Euler, expressing sincere regret about Euler's illness and the loss of the use of his eye, and asking for more specific information on the nature of the infection:

> Tell me, was your eye completely destroyed or is the eyeball still intact? Did the fluid inside the eyeball leak out?
>
> I guessed that you had not told your parents about your illness this time either, and when I went to talk with them, I found that I was right. They knew nothing about it. They are, of course, concerned but realize that there is little that they can do. However, I hope you will write to them soon so that they have the reassurance of hearing directly from you. As you are undoubtedly aware, they are both extremely fond and extremely proud of you.

That is the kind of affection that Daniel Bernoulli never felt from his own father.

Euler's reply is not part of the historical record, although later pictures of Euler show nothing worse than a drooping of the right eyelid—there is no gaping eye socket.

Euler blamed the infection and the loss of his eyesight on some laborious calculations on which he had spent three intense days. Since the calculation was urgent, quite naturally Euler had to do it since it would have taken anyone else several months. It was simply a fact that Euler was the fastest calculator in the world. A friend described him as someone who calculates as easily "as men breathe or as eagles sustain themselves in the wind."

However, even Euler had his limits. One day when he sat down to write to Goldbach, Euler had received another series of maps that had been sent to him for his careful examination. He wrote to his friend and permanent secretary of the academy:

> My Dear Sir!
>
> Just today I have received a packet of maps that I am supposed to study carefully and use to make recommendations for the government. I have a huge favor to ask of you, dear sir. I think you are aware that I have lost the use of my right eye after a serious infection that followed several days of intensive calculations relating to mapping. Could you please use your influence to attempt to remove geography from my duties? I think such calculating is responsible for the loss of my sight in one eye, and although I can get along with only one eye, I seriously do not want to lose the sight in my remaining eye. Please sir, if you value me as a friend and a scientist, try to help me in this. Geography is fatal for me!
> Your, Euler

It appears that Goldbach unfortunately was not able to arrange for Euler's release from his geography responsibilities. While it is known today that stress does little more than act as an initial impulse for a physical malady, at the time medical experts confidently blamed it for the whole event. Euler considered the serious study of a large map more taxing on his eyes than simple reading and writing and calculating because it forced him to study a larger area while at the same time demanding that he focus on the details within it.

Pragmatist that he was, Euler then went back to work with his usual vigor. He even commented to someone who was commiserating with him on the loss of the use of his eye, "but now I shall have less distraction!" He maintained that with one eye he could pursue his work just as well as with two. Euler was a realist who accepted his lot without complaint.

33
St. Petersburg Loses Euler to Frederick the Great of Prussia

After Anna Ivanovna, Tsarina of Russia, died in 1740, Euler found the political climate in St. Petersburg increasingly uncomfortable. The infant Ivan VI's ascent to the Russian throne (whose horoscope Euler had been asked to cast but instead quietly passed on to the court astrologer) provided yet another occasion for a power vacuum with yet another difficult regency. Once again, relatives in the royal family viciously competed with one another for power. Within a year, a palace revolt placed Peter the Great's daughter, Elisavetta Petrovna, on the throne, where she would remain for 20 years. When she assumed power, the atmosphere within the academy became less and less congenial.

The government of the new Tsarina Elisavetta, along with Russian society in general, resented foreigners in their midst, and the academy, in which most of the scholars were obviously foreign, was a prime example of such undesirable non-Russian influence. Even though Euler alone among the foreign scientists spoke fluent Russian, he was still a foreigner and thus suspect. At this time, however, Euler got some welcome news: he received word from Berlin that Frederick the Great, King of Prussia, was eager to lure him to his own academy that he had decided to revive in Berlin. Euler saw this as a positive development for several reasons.

Although mild-mannered Euler probably could have continued to survive and work productively in the St. Petersburg Academy, he nevertheless believed that his work in geography, which was still a major part of his assignment at the academy, seriously threatened his vision. He would later write about the miraculous function of the human eye, "the most marvelous thing that the human spirit can penetrate." As a pious man, he saw the human eye as the most perfect example of the infinite wisdom of God, and he didn't want to risk losing the use of his remaining eye. As an active scientist, he lived in dread of a state of total blindness.

His wife Katharina was also urging him once again to find a way for them to leave St. Petersburg, not least because of her dread that a fire might at any time consume their house, which she often reminded him was made of wood. Katharina had always kept packed suitcases ready in case the family should need to escape from a fire in a hurry. Her fear was justified since houses in the city of St. Petersburg often caught fire, causing their occupants to lose everything. Because all the homes were heated and lit with open fires in fireplaces and with candles, the danger was constant, and a fire in one house could easily cause rampant fires throughout a large part of the city. In May of 1771, when the Eulers were once again living in St. Petersburg, Katharina's greatest fear was realized as 550 houses, including the Eulers', burned to the ground. Katharina assumed correctly that if they were to move to Berlin, they would be able to purchase a relatively fireproof stone house. Although Euler had been active in developing better tools for the St. Petersburg fire brigade, once a house caught fire, there was little hope of saving it.

Another complaint the Eulers had about life in St. Petersburg was the constant threat of being forced to quarter soldiers in their house, because of a perpetual shortage of housing in the city. Euler and his wife resented the sudden arrival of an uncouth soldier into their family circle, which by then was growing large.

"Leonhard," Katharina said to her husband one evening as the family was gathering for the evening Bible reading and prayers, "the colonel informed me this afternoon that we will have to quarter three soldiers in our house for two weeks, beginning this evening."

"That is outrageous!" Euler said. "With our growing family, we simply don't have enough room in our home to accommodate them. What did you tell the colonel?"

"I'm afraid we don't have a choice," Katharina said. "I told him we would do what is required. I have moved the boys into our room so that the soldiers can have the use of their room."

"I wish that were not necessary," Euler said grimly. Then carefully considering the situation, he philosophically picked up his Bible and opened it in the New Testament to the beginning of Hebrews 12. He read, "Wherefore seeing we also are compassed about with so great a cloud of witnesses, let us lay aside every weight, and the sin which doth so easily beset us, and let us run with patience the race that is set before us."

Just then there was a sharp knock at the door. Euler, glancing unhappily at his wife, went to open it, and found four—not just three—soldiers arriving with their packs and weapons. He greeted them civilly in Russian, "Yes, you are the soldiers who will be staying in our house, I presume."

"Yes, Professor Euler," one of the soldiers replied, demonstrating that he understood the importance of his host—St. Petersburg's most esteemed scholar. "If we could just put our gear in the room where we will be staying, we can then go out and join the rest of our company for drill and then our evening meal."

By now Katharina had joined Euler at the door. "I can show them in, Leonhard," she said quietly to her husband. Then to the soldiers, she said, "Please follow me. I'll show you to the room where you will be staying. Leonhard, there is no need for you to wait for me to continue your reading. The children and the servants are waiting. I will be back downstairs in just a few minutes."

Euler then returned to his seat and explained, "You see, my children, we are not promised a life without difficulties. We must learn patience." Euler glanced around the room, at the family who were gathered for the evening lesson. Then he spoke again, "Now Hebrews continues in chapter 13, 'Let brotherly love continue. Be not forgetful to entertain strangers: for thereby some have entertained angels unawares.'"

The family then became aware of the sounds of heavy boots in the hall. "We'll be off now," one of the soldiers announced loudly from the hall as Katharina quietly sat down next to her husband once again.

Euler raised his voice to address the soldiers, "When you come in tonight, please be careful to secure the bar on the front door."

"Yes, Professor," one of the soldiers replied. "We will certainly do that. Good evening." With relief, the family heard the front door close behind the departing soldiers.

As calmly as he could, Euler returned his attention to his family, saying, "In verse 6 of chapter 13, the Scripture says 'The Lord is my helper, and I will not fear what man shall do unto me.' And verse 16 continues, 'But to do good and to communicate forget not: for with such sacrifices, God is well pleased.'" Closing his Bible, Euler said, "We have not chosen to provide housing for those soldiers, but neither have they chosen to lodge with us. We must assume that they are fine young men, and we must treat them with that in mind. I may fuss about their intrusion into our family life, but we must always treat them with courtesy and charity. Indeed, it is possible that one of them may be an angel in our midst, as the Scripture says. Never forget that lesson, my children. Amen."

Seeing that her husband had completed the evening lesson, Katharina rose and gathered the children and quietly took them upstairs to bed. Euler sat still for a few minutes, contemplating his current burden. "Dear Lord, please give me the patience to continue in my work and our family life in spite of the presence of these rude soldiers, who certainly would not have chosen to interrupt our family circle. With your guidance, we will try to retain our Christian mission in this less than perfect world. Amen."

When he received an official invitation to the court of Frederick the Great, Euler made a formal request to Schumacher to allow him to depart from Russia. Such permission, however, was not automatic. During his years in St. Petersburg before 1740, Euler published more than 50 essays and books, covering a wide range of subjects. He wrote works in algebra, number theory, Diophantine analysis, geometry, topology, differential geometry, differential equations, infinite series, calculus of variation, mechanics, theory of ships, physics, astronomy, the theory of tides, and music theory. Euler, who had made major discoveries in each of the fields, was the expert in all of them. By this time, he was universally recognized as the most important mathematician of his time.

Although the new regime in Russia was making life unpleasant for the Eulers, in spite of himself Schumacher realized that if Euler were to leave, it would be a great loss to the academy. He might not like Euler personally and he might persecute him in a thousand little ways, but he knew Euler's reputation, and he was reluctant to lose him. How was the academy at St. Petersburg to compete with Frederick the Great's academy in Prussia without the services of Euler, particularly if Euler were to move to Frederick's side in the competition? With that in mind, he simply refused to allow Euler to break his contract with the academy. Euler could not depart without Schumacher's permission, despite the generous and desirable invitation from Frederick the Great.

In the end, the Prussian ambassador to St. Petersburg arranged for Euler to be diagnosed as ill, so that he could finally be released from his contract for medical reasons. As it turned out, Schumacher eventually realized that Euler's departure from St. Petersburg did little to hurt the academy's reputation. After he left, Euler continued to submit hundreds of essays each year to be included in the journal of the Russian Academy of Sciences regardless of the fact that he was no longer living in St. Petersburg. One academic press simply couldn't keep up with Euler's remarkable output.

When Euler's 73-year-old mentor Johann Bernoulli heard of Euler's plans to move to Berlin, he was delighted, immediately suggesting that now Euler could come to Basel and visit his parents, whom he hadn't seen since he was 20 years old. Euler would no longer be at the very edge of civilization as Johann Bernoulli saw it, and he hoped his most famous student would find the time to make the trip home. A further selfish implication was obvious—a visit to Basel to visit his parents would also allow Euler to see his mentor Bernoulli, who had been eagerly following his prodigy's career from a distance for many years.

In a second letter, Johann also inquired about the salary Euler had been offered in Berlin, clearly wondering if Frederick might be willing to pay one or both of his sons, Daniel or Johann, well enough to entice them also to move to Berlin. The elder Bernoulli admitted in his letter that if he were 20 years younger, he would jump at the opportunities available in Berlin, but at his age it made no sense. Euler, who was always busy with many disparate projects, never found the time to make that trip back to Basel, and he never again saw either his father or the elder Johann Bernoulli. Daniel and the younger Johann, who never accepted positions in Berlin, occasionally visited Euler there.

34
The Eulers Arrive at the Court of Frederick the Great in Berlin

Frederick the Great, who considered himself the most important and most enlightened ruler in Europe, was enthusiastic about accumulating an impressive collection of scholars to stock his academy. Only the best were good enough for Frederick. Since Euler was clearly the most brilliant and most productive mathematician in the world, he was a natural choice. Frederick's offer to Euler promised a salary of 1,600 Taler per year in addition to 500 Taler to cover his moving expenses. This was a generous offer, which Euler was able to supplement with an additional stipend from St. Petersburg as well, since the academy there still continued to benefit from and was willing to pay for the works of this amazingly productive scholar. Not only did he continue to write many of the articles that were included in this journal; he also continued to be the general editor in charge of the journal.

However, it didn't take long for Euler to realize that the situation in Berlin was not ideal. There was never any doubt about what kind of scholar the Prussian King Frederick preferred—a scintillating French savant—and Euler didn't qualify. Frederick the Great had been brought up speaking French, rather than German. Having the greatest admiration for all things French, he desperately wanted the witty and urbane Voltaire as the crowning jewel in his academy. Unfortunately for Frederick, however, Voltaire had the audacity to refuse. He had a cozy living arrangement with the mathematician the Marquise Emilie du Châtelet at her manor in Cirey, France, where the two of them talked and studied and worked together exactly as they pleased. Frederick the Great might be an interesting host now and then, but Voltaire and his Marquise had the life that they wanted without any need to compromise their freedom. In the process of wooing him, Frederick had written to Voltaire that he expected to run his academy, of which Voltaire would be president, in the same way that a squire keeps a pack of dogs. Voltaire, who had never been one to toady to any authority figure, had no intention of being a dog—top dog or not—in any king's kennel. Frederick, disappointed at his dealings with Voltaire, was forced to settle for Euler, who spoke French easily and beautifully but with perhaps the suggestion of a Swiss accent. Frederick knew Euler's German-speaking background, and he found it wanting.

In spite of his disappointment, Frederick knew that the humble Euler was essential to the prestige of his academy. He was pleased that he had been able to attract Euler to Berlin, and he was happy to take advantage of that scientist's universal genius. Frederick, who was well aware that Euler preferred the German language over French, dismissed Euler as something of a country bumpkin. Euler was a pious man, not impressed by the silly emphasis in Frederick's court on elegant manners. Euler preferred solid, honest, intellectual discourse. Frederick, for his part, found his new scholar simply ponderous.

When the Euler family arrived in Berlin on July 25, 1741, Euler, who was then 34 years old, made arrangements to buy a large, stone, presumably fire-resistant house on Behrenstrasse in central Berlin. Although it took a year to complete renovations and repairs, when it was finally ready, the family of seven was delighted with it and lived there comfortably for 23 years. With the suitcases conveniently stashed in the attic, Katharina Euler could finally sleep through the night in her stone house with little worry that it would catch fire.

Plaque on Euler's house in Berlin: Here lived the Mathematician Leonhard Euler from 1743 to 1766; 15 April 1707–18 September 1783. In his memory, the city of Berlin, 1907.

In early September 1741, during the War of the Austrian Succession, and three long months after his arrival in Berlin, Euler finally received his first letter from Frederick the Great, written in French from a military camp.

> *Monsieur* Euler,
>
> I was pleased to learn that you are happy with your present position. I have given the necessary orders to the Director for a salary of 1600 ecus that I offered you. If there is anything else that you need, you have only to wait for my return to Berlin.
>
> I am your affectionate king, Frederick

In fact, Euler had to wait at least three more months to receive any of the promised money, during which time he was forced to live on credit. It would be far longer before he actually met the king. A year later, in September of 1742, the Royal Brandenburg Society formally welcomed "the famous professor of mathematics, Mr. Euler."

Euler was also soon formally presented to Frederick's mother, who prided herself on being a charming hostess. Although Euler was certainly a charming man in his own quiet way, he did not have the polished manners of a French nobleman—the manners that were valued in the sparkling court in Berlin.

"Good evening, *Monsieur* Euler," she began in French, "I hope you are finding the arrangements here at the academy in Berlin satisfactory."

"Yes, *Madame*," Euler replied.

"And have you found that the other scientists in the academy are congenial?" she asked.

"Yes, *Madame*," Euler replied.

"My son, his majesty King Frederick II, has asked me to tell you that if there is anything else that you need, you should please just let me know," she said.

"Thank you, *Madame*," Euler said.

"*Monsieur* Euler," she said in consternation, "can you say nothing beyond a two or three word answer? Can't you see that I am trying to get to know you? I would like to have a conversation with you."

"But *Madame*," Euler protested, "you may find this hard to understand, but where I have been living in imperial Russia, if a person speaks out, he will be hanged! Of necessity, I have learned to be cautious, *Madame*."

That got the queen mother's attention. Although she might not like it, she had to accept the fact that it was probably literally true. She was forced to resign herself to the fact that the Eulers were not going to fit easily into life at the fashionable Prussian court in Berlin.

Euler had assumed that he would be chosen as either the head of Frederick's academy, or at the very least as the head of the mathematics section, but Frederick had other ideas. Failing to get Voltaire, he had chosen Maupertuis, the celebrated traveler on the polar expedition, whom young Johann Bernoulli had described to him and Daniel in Berlin. Maupertuis, by now an accomplished mathematician (thanks to his mentor Johann Bernoulli's help) and always a brilliant conversationalist, became Frederick's scholar of choice. In January 1746, Frederick named Maupertuis president of his academy at a generous salary of 3,000 Taler and absolute power over everything within the academy. At the time, Maupertuis did not know Euler, but he had certainly heard about him from the Bernoullis. The arrogant Maupertuis, who was no fool, agreed with the modest Euler on many questions of science, while at the same time he relished his total power over Euler in the academy. Their working relationship was not always easy for Euler.

At this time in his life, Maupertuis, who was suffering from what has been described as a "lung disease," was not particularly happy in Frederick's Prussia. He preferred to stay in what he considered the healthier climate of France. In fact, for many years Euler served as the de facto head of the academy although he never was given the status of official head or the salary to match it. Maupertuis recommended to Frederick that Euler be officially asked to serve as acting president of the academy from 1753 to 1754, justifying his recommendation with Euler's many obvious talents. Frederick agreed, but with little enthusiasm. Regardless, Maupertuis still kept close track of what was happening in the academy and gave Euler little real freedom to make the decisions that he considered necessary.

When Maupertuis died in the home of young Johann Bernoulli in Basel in 1759, Euler, who was then 52 years old, once again served as acting president of the academy. Euler hoped Frederick would make the obvious move this time—make him president of the academy—but over the years Frederick had come to actively dislike Euler. Frederick wanted the Frenchman Jean d'Alembert, a recognized mathematician who was active in the Paris Academy, to serve as the new president of his academy instead. Six years earlier, Euler had first heard of d'Alembert through his friend Daniel Bernoulli, who had contempt for d'Alembert and his mathematics. Euler was inclined to trust Daniel's opinion.

In fact, Frederick soon offered the presidency of his academy to d'Alembert, who rejected it after a three-month visit to the academy in Berlin in 1764. Although Frederick was convinced that his court was a French court, it didn't fool a real Frenchman. Berlin was Berlin—the center of German culture—and no one would ever mistake it for Paris.

Euler was surprised later to learn that part of d'Alembert's reason for turning down Frederick's offer was that he could see that Euler was a far better mathematician than he was. After turning down another generous offer, this time from the St. Petersburg Academy, d'Alembert eventually became perpetual secretary of the French Academy, the academy that he preferred above all others. From this time on, however, Frederick actually treated d'Alembert as the secret president of his academy anyway, allowing him to carry out his duties by correspondence, and once more undermining Euler's position.

The academy, which was always under-funded, had the sale of almanacs as its main source of income, and their production demanded significant work from Euler.

"Sire," Euler addressed Frederick, "I was hopeful that the sale of almanacs this year would produce a profit of 16,000 Taler, but I now see that we will be 3,000 Taler short of that this year. Next year's almanac should be closer to my goal."

"*Monsieur le professeur*," Frederick sneered, "your calculations are impossibly complicated. If you would just use standard methods I am sure you could produce a perfectly satisfactory almanac which might even make some money."

"No, Sire," Euler argued, "I have found that the calculations in the almanacs from the past few years are terribly inadequate. Simple arithmetic is not enough for these calculations."

Almanac for the year 1753.

"It always was before," Frederick argued.

"Regardless of whether anyone complained to you about the earlier product," Euler explained, "it had great potential for throwing off a farmer's plans. You may not like my mathematics, Sire, but the almanac under my direction is now respected as the only acceptable calendar for a farmer to use."

"You are 3,000 Taler short!" Frederick complained. "That is arithmetic I can understand."

Like Benjamin Franklin's *Poor Richard's Almanac* in Philadelphia in America, Euler's almanacs adorned the kitchen table of any resident of Prussia who could possibly find the money to pay for one. The farmers who depended on the almanacs benefited day after day from the fruits of Euler's labors.

Not only were Euler's almanacs full of practical information—and the only source of such information in a common farmhouse—they were also charming. A typical almanac (see illustration) was about three inches by five inches and perhaps one inch thick. The right-hand page for the month of August 1753 has a picture of a man harvesting hay with a scythe (see illustration), with this poem below the picture:

> See how diligent is the farmer,
> Though tired and sweaty, he still works with ardor.
> The scythe tires him, though he sees its blessings.
> May we always remember hard work's lessons.

Above the picture in the upper left and right corners, the calendar helpfully tells us, "August has 31 Days."

The left column of the left-hand page gives the date and the day of the week for each day, followed by a name, such as Gustavus or Bertram. The right column gives such information as general

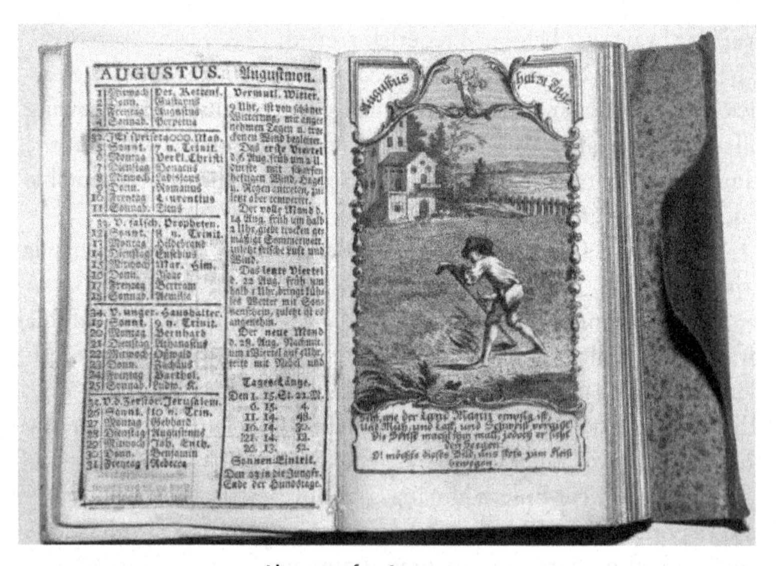
AUGUSTUS. Augustmon.

Almanac for August, 1753.

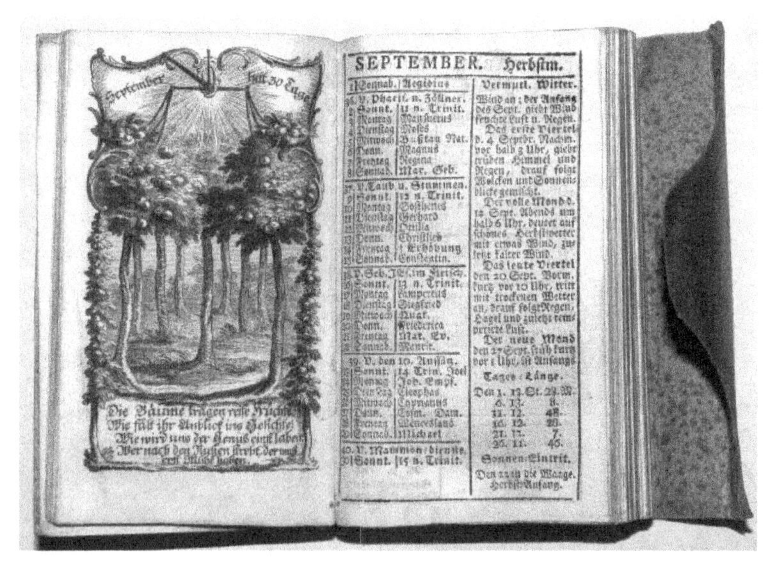
SEPTEMBER. Herbstm.

Almanac for September, 1753

weather forecasts, the date of the new moon that month, and the time of sunrise and sunset on selected days.

September's picture shows apple trees covered with many ripe apples, and its poem says,

> The trees are so heavy with fruit!
> Oh! Those apples look to us so good!
> Oh, what a beautiful sight!
> What a great reward we get for using our might!

Once again, September comes with important information on the weather, the date of the new moon, sample times of sunrise and sunset, and the important information that this year fall will begin on September 22.

Site of the estate that Euler bought for his mother in Charlottenburg.

In 1750, Euler's only brother Heinrich died in Basel. He had returned to Basel several years earlier from St. Petersburg and Paris and had pursued a successful career there as an artist. Now that his mother was alone since his father had died in 1745, Euler urged her to come live with him in Berlin. Since she had always lived in the quiet village of Riehen, she agreed but expressed reluctance to live in the busy city of Berlin. In 1753, Euler found a lovely estate for her in the suburb of Charlottenburg costing 6,000 Taler, not far from the palaces constructed by King Frederick's grandfather and named for his grandmother Queen Sophie Charlotte.

Euler traveled to Frankfurt in the company of his wife Katharina and their oldest son Johann Albrecht to meet his mother and convey her to Berlin. His mother was delighted with the estate (see illustration), where she lived for several years, enjoying the company of many of her grandchildren and supervising a large and successful farm there. After a long day at the academy, Euler often enjoyed walking the mile to visit his mother at her estate.

Over time, Frederick the Great became even more irritated by Euler's less polished ways, and this continued to make Euler's time in Berlin difficult. Frederick relished openly referring to Euler as his mathematical cyclops, a cruel reference to Euler's loss of vision in his right eye. On the same theme, Frederick once wrote in a letter to Maupertuis that he would prefer to have a mathematician with two eyes, not just one. Euler, with his simple country ways and his devout lifestyle, had none of the glitter that Frederick expected in his resident scholars.

At this time, Frederick the Great was engaged in the Seven Years War with Russia and Austria. Although Frederick may have disliked Euler the mathematician heartily by this time, one of Euler's invaluable functions for him was to translate intercepted messages from the Russian military. Euler was the only member of Frederick's court who could speak and read Russian. Frederick was well aware that Euler was a man of many talents, and Frederick didn't hesitate to take advantage of him when it was convenient.

When the Russians temporarily occupied Berlin, the general in charge instructed his commanders that Euler and his properties were not to be damaged, but somehow Euler's mother's estate in Charlottenburg was destroyed anyway. Fortunately none of his family was injured during the military operation. Learning what had happened, the general immediately apologized profusely to Euler and paid him in full for all his losses, saying that he was waging war with Frederick, not with the sciences. When the Empress Elisavetta back in Russia heard of the incident, she sent Euler 4,000 Rubles more. Euler was to be respected at all cost.

On top of all the creative work that Euler did in Berlin, he enjoyed some leisure time as well. Playing the clavier was always one of his pleasures, and he enjoyed entertaining musicians in his home, watching and listening as they played. Although he often dealt with music as science, he was not deaf to its aesthetic charms.

Euler also spent some time playing the game of chess, which he was reasonably good at. He studied with a chess master in Berlin, developing enough skill at the game to occasionally challenge the master himself seriously, although Euler never claimed to be a chess master. He also enjoyed reading to his children and grandchildren, and was particularly fond of taking them to the Berlin zoo to watch the bear cubs play.

35
Euler's Scientific Work in Berlin

While he was in Berlin, Euler was as creative and diligent as always. In addition to the almanacs, he had many other special assignments from the king. For example, Frederick asked him to design pumps for the fountains in the gardens at his lovely new palace called *Sans Souci* [Without Cares] in Potsdam, outside of Berlin. In the end, Frederick was not satisfied with the results, because Euler reported that the current wooden pipes reinforced with metal couldn't contain the water pressure that Frederick wanted. Euler had his limits—even he couldn't defy the laws of physics.

On the plus side, Euler satisfied Frederick when he masterminded the construction of the Finow canal, which connected the Oder and the Havel rivers, using a total of 17 locks. As a result, the landlocked city of Stettin became a viable maritime port. Euler was also involved in the planning of water and windmills as well as assessing the lottery systems then in use in Italy and Holland, with an eye to creating a lottery system in Prussia—a tantalizingly easy way to bring in additional money for the state. Euler was careful not to include in his report his own disapproval of such an unfair way for the kingdom to raise money.

However, in addition to all his practical work for Frederick between 1741 and 1766, the scientist and scholar Euler also wrote more than 380 essays and books on a wide variety of topics. One such writing project for Frederick the Great, the soldier, was a translation

Frederick's palace Sans Souci in Potsdam, Germany.

of Benjamin Robins' 1742 book on gunnery, a work that was largely unknown, but that the Prussian king needed for his army. Until this time, the only book on the physics of gunnery had been written by Galileo in the early 1600s. Although Galileo had begun the study of gunnery, he didn't have the mathematical or physical tools that would have allowed him to factor in the effect of air resistance. By Robins' and Euler's time, air resistance had become a serious consideration, and physicists now had Leibniz's calculus, which allowed them to deal with it. Robins' book presented formulas and tables that improved dramatically on Galileo's methods.

Since Robins' book was in English, Euler's first task was to translate it. In addition to his fluency in German, French, Russian, and Latin, apparently by this time Euler was able to read and translate English as well, although he had never traveled to England and probably couldn't actually speak English very well. For Frederick's purposes, the book needed to be in German so that his common soldiers and officers could read it. Although the monarch preferred French, his soldiers certainly spoke only German. If Euler had translated Robins' book into Latin, which he could easily have done, only scholars could have read it—not the intended audience. Euler's German translation was soon translated again into French, resulting in a version that the French army was still using during the Napoleonic wars and even as late as the beginning of the First World War.

The modern edition of Robins' book in English has 190 pages; Euler's translation and further clarifications (of which there were many) produced a book whose modern edition has 514 pages. Robins' work, published in 1742, was certainly a major accomplishment, including tables that allowed a soldier in the field to quickly find the exact angle at which he should set his cannon in order to aim it accurately at a target. From Frederick the Great's viewpoint, Euler's edition of Robins' work with its many additional tables and calculations, which came out only three years later in 1745, was probably Euler's most important work. In fact, it was only after Euler published his translation that Robins' work attracted any notice, so that Robins was in Euler's debt for any recognition that he got for the work.

Another project that Euler undertook for Frederick was a work that was originally published in French (since the language of the Prussian court was French) under the title, *Lettres à une Princesse d'Allemagne* [*Letters to a German Princess*]. It was a series of letters to the 15-year-old princess Sophie Friederike Charlotte Leopoldine von Brandenburg-Schwedt, one of Frederick's nieces, who would spend most of her life at a convent for noble ladies, where she was made abbess in 1765. The letters provide instruction on a remarkable range of topics from geometry and music theory to philosophy, mechanics, optics, astronomy, theology, and ethics. It appears that the instruction of the princess had begun in person, since the first letter suggests that it is a continuation of her instruction in geometry. Euler always addresses her as *V.A.*, an abbreviation for the French *Votre Altesse* [Your Highness].

Euler was sometimes criticized for his writings on philosophy and ethics, since those were not subjects that he had studied formally.

In fact, his friend Daniel Bernoulli once told him frankly, "you really shouldn't write about these topics about which your knowledge is not complete, since people expect only the most sublime writing from you."

The *Lettres* were soon translated into German, English, Russian, Dutch, Swedish, Danish, Italian, and Spanish, becoming a bestseller in all these languages as well as French. Euler wrote the 234 letters between 1760 and 1762, and they were later published in St. Petersburg in three volumes, beginning in 1768. The work, which is still available, is one of the clearest explanations of popular science written for a nonscientific audience that has ever been written. It is remarkable that it was the work of the foremost mathematician and scientist of his day who was regarded as the most erudite man in the world. However, he demonstrated once again that although he could explain clearly the most abstruse and revolutionary concepts in his scientific works for a knowledgeable audience, he could also explain basic science to a general audience.

Beyond Frederick's specific assignments, another topic that Euler explored at this time was what we now call graph theory. Euler was intrigued by the fact that a polyhedron with straight edges and flat surfaces always displays an interesting feature: the sum of the number of vertices plus the number of faces is always equal to the number of edges plus two. This is sometimes called Euler's formula, which can be written algebraically as $V + F = E + 2$. It is true of a cube where 8 vertices + 6 faces = 12 edges + 2, and it is true of any other convex polyhedron. The question is not whether it is true—clearly it is. What Euler needed to do was to prove that it is always true, and that is difficult. His first proof, published in 1750, was flawed, but his second proof, published in 1751, although still not perfect, was pretty good.

Unlike many great scientists, Euler was always generous when it came to questions of priority, such as the nasty calculus wars in which his mentor Johann Bernoulli engaged. When in 1755 the 19-year-old French mathematician Joseph Louis Lagrange (1736–

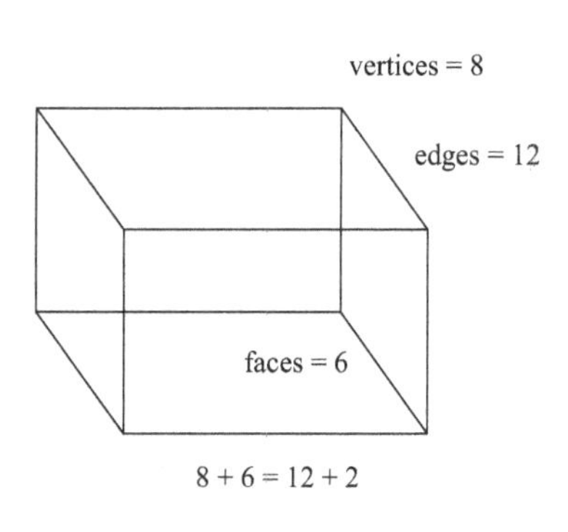

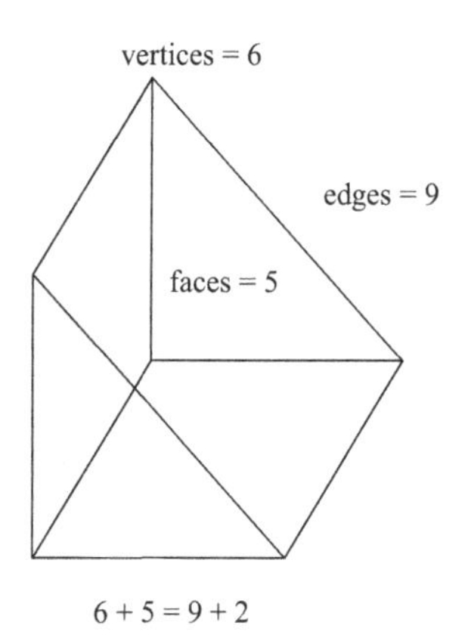

Euler's famous formula: $V + F = E + 2$ where V = number of vertices, F = number of faces, and E = number of edges.

1813) wrote to Euler, who was then 48 years old, about a discovery he had made in solving isoperimetric problems (see Dido's problem in Chapter 15), Euler read Lagrange's letter and agreed that Lagrange had succeeded in a solution that Euler himself had not quite reached, although he had been working on it for some time. Instead of quickly publishing his own work on the subject, Euler deliberately held off so that Lagrange could publish first and be given credit for his discovery. Euler deferred to Lagrange on several of his other discoveries as well—for example Lagrange's discovery that every natural number can be written as the sum of four squares—because he saw that Lagrange's treatment was more complete than his own. Euler was pleased to see a young mathematician make a good start in his career, and he recognized beautiful mathematics when he saw it, no matter who discovered it. In fact, a few more brilliant discoveries added to Euler's phenomenal output would have made very little difference.

Euler's 1748 textbook called the *Introductio in Analysin Infinitorum* [*Introduction to Analysis of the Infinite*] was the first complete textbook in analysis. Like his other textbooks, it was written in clear Latin using modern notation, so that a student anywhere in Europe at the time, and even today, can read it easily in the original Latin or in translation. It is such a classic that today it is often referred to with its shortened Latin name as simply the *Introductio*—everyone knows what it is an introduction to! In writing it, Euler brilliantly determined the structure of the entire field. His revolutionary approach seems today like the obvious way to do it, but that is only because Euler gave it to us. It represents the first exposition of function theory, and with it Euler established analysis as the science of functions—statements that relate one quantity to another. Because of Euler's text, we now write functions in the way that Euler learned from Johann Bernoulli, who learned it from his brother Jacob: $f(x)$ is read as "a function of x."

In the *Introductio*, Euler was careful to limit himself to pre-calculus material, allowing his readers to master the basics of functions as they relate to algebra, geometry, and trigonometry, before leading them to the realm of the calculus. Although he was a scientist working in the forefront of his many fields, he still had a clear sense of all the steps necessary to construct students' understanding of the foundations of higher mathematics. He considered higher mathematics fully as important to the development of an educated mind as arithmetic, algebra, and geometry, in stark contrast to many scholars and laymen of his day as well as our own—including Frederick the Great.

Euler's 1755 basic textbook in the differential calculus followed his *Introductio*, providing the first complete treatment of that subject. Once again, he clearly presented the structure of the subject to his readers. After defining the differential calculus, he then proceeded to explain its concepts clearly, although without the use of diagrams. His was a text in pure mathematics—not physics. He was certainly aware of the many applications of differential calculus to real-world problems, but in this book—a work of pure analysis—he deliberately confined himself to the basic concepts.

Euler's 1768 textbook on integral calculus, which was published after his return to St. Petersburg, followed Johann Bernoulli's plan of presenting integration of an expression as the inverse of its differentiation. After presenting basic techniques for integrating, he then moved on to the summation of infinite series.

These three works that Euler wrote on pre-calculus and calculus were the most influential textbooks on these subjects in the eighteenth century. Although we have no records of how many copies were printed or sold, we know their effect. The fact that the notation that he presented in his books is the notation that has survived is clear proof of their influence. Mathematicians and scientists in England soon adopted Euler's notation as well, completing his mentor Johann Bernoulli's conquest of Newton's fluxions, which was probably not Euler's goal.

The works discussed here are only the beginning of Euler's list. During his adult life, Euler composed an astonishing 800 pages of original mathematics every year for 60 years! His most famous formula, which he didn't actually write in its present form, involves the five important constants in mathematics: $e^{\pi i} + 1 = 0$, where e is the base of the common logarithms (2.71...), π is the ratio of a circle's circumference to its diameter, i is defined as $\sqrt{-1}$, and zero and one are the foundations of the number system from which all other numbers can be constructed. It is mind-boggling that one mind could have produced and presented such an amazing variety of important scientific work, important to the layman as well as to his fellow scientists.

36
Euler's Work in Number Theory

After he had arrived in St. Petersburg in 1727, then 20-year-old Euler had come to enjoy the company of Christian Goldbach, a friend of the Bernoulli family who traveled actively throughout Europe and was a key player in the academy at St. Petersburg for many years. Although Goldbach was not a mathematician, he enjoyed playing with the subject and was well enough informed to carry on serious conversations on many of its topics. He also was in frequent communication with many of the amateur mathematicians of Europe at the time.

In October 1729, 22-year-old Euler closed a letter that he wrote to Goldbach in Latin, describing some of his recent work on infinite series, addressing Goldbach as "*Tu, Vir Celeberime...* [you, most celebrated man..., who has studied such series, will be interested in the way that this develops]." In his response, Goldbach addressed Euler as *Vir Clarissime* [brightest man], responding to Euler's work on the series in his letter. After signing off, Goldbach added a now famous postscript: He asked, "What do you think about Fermat's primes?"

Goldbach was referring to Pierre de Fermat (1601–1665), a judge in southern France in the previous century, who chose to spend his leisure time playing with mathematics, specifically number theory, rather than passing time in the company of his fellow citizens, since at any time he might be called upon to sit in judgment of them and he didn't want to risk any personal prejudices. Fermat had studied some mathematics in the process of preparing for his career in law, but most of what he accomplished in mathematics, he did on his own. He corresponded with other serious mathematicians of his time like Marin Mersenne (1588–1648), René Descartes (1596–1650), Blaise Pascal (1623–1662), etc., and seemed to take particular pleasure in tossing out challenges to his correspondents. Very little of his work was published until after his death, and many questions remain concerning his proofs, which he rarely included in his writings. Eric Temple Bell in his 1937 classic book *Men of Mathematics* described Fermat as "the prince of amateurs."

Fermat's most famous comment (called Fermat's Last Theorem today) is found in his marginal notes to a Latin translation of Diophantus' (A.D. 250) works, where Fermat wrote that the expression $x^n + y^n = z^n$ is true for no integer n that is greater than two. There are many sets of numbers x, y, and z for which $x^2 + y^2 = z^2$ (the Pythagorean Theorem) is true—for example $3^2 + 4^2 = 5^2$ and $5^2 + 12^2 = 13^2$. Fermat's statement was this: There are no whole numbers x, y, and z such that when all three numbers are raised to the same power that is greater than two, the equation is true. For example, there are no integers x, y, and z for which $x^3 + y^3 = z^3$. He then famously wrote, "I have a marvelous proof of this, but this margin is too small to include it." Mathematicians tried to discover his proof (or any other proof, for that matter) for the next 350 years. Andrew Wiles at Princeton University finally accomplished it in 1996, leaning heavily on several discoveries that had been found in the intervening years. It is likely that Fermat's own proof was flawed although his conclusion was correct: There are no values of x, y, and z for which $x^n + y^n = z^n$ is true for any value of n greater than two. Since we don't have Fermat's work, we can only be skeptical about the validity of his proof.

Goldbach's postscript to his letter to Euler concerned another question posed by Fermat: When you raise the number two to a power which is of a power of two and then add one, does it always result in a prime number? Fermat thought it did, but he didn't claim to have a proof. Goldbach wondered if Euler had any thoughts on the question.

Euler, who was working diligently on many different parts of mathematics at the time, was probably surprised by Goldbach's question. Was he suggesting that Euler should waste his time on number theory? Goldbach and his friends apparently enjoyed playing with this sort of popular mathematics, but in the scholarly world the subject was regarded as trivial. Goldbach should have known that work in that field would do little to advance Euler's career, which was then just starting. However, even though Euler knew it was not wise for him to get sidetracked by this silly question, in fact he could not resist it.

Here is the challenge as Goldbach quoted Fermat's algebraic question: For the expression $2^{2^{x-1}}+1$ where x is any whole number, is the resulting number guaranteed to be a prime number? Euler quickly constructed the first few Fermat Primes. "If I take $x = 1$, it gives me $2^{2^{1-1}} + 1$, that gives me $2^{2^0} + 1$, or $2^1 + 1 = 2 + 1 = 3$, which is certainly a prime number. If I take $x = 2$, I get $2^{2^{2-1}} + 1 = 2^{2^1} + 1$ which gives me $2^2 + 1$, in other words $4 + 1 = 5$, another prime number. When $x = 3$, I get $2^{2^{3-1}} + 1$ which gives me $2^{2^2} + 1$ or $2^4 + 1$ or 17, another prime number." This quick exploration took Euler only a few minutes.

In the postscript to his letter of January 1730, Euler put Goldbach off:

> I can come up with nothing absolutely to prove Fermat's statement, but I am not yet persuaded that it can be proven with certainty.... I'll let you know if I find a proof. I ask you to wait, esteemed gentleman.
>
> I am indebted to you, Euler.

Over the next few weeks, Euler spent a few idle moments considering Goldbach's question. In a May 1730 letter he was pleased to have an answer for Goldbach, although by this time he had decided to change the form of the problem, making it $2^{2^x} + 1$, a simpler form of the expression that would lead to the same results. While playing around with his new version of Fermat's statement, he tried $x = 3$ in his formula and got the next Fermat prime: $2^{2^3} + 1 = 2^8 + 1 = 256 + 1 = 257$, a number that Euler recognized as prime. Euler also noted that when $x = 4$, then $2^{2^4} + 1 = 2^{16} + 1 = 65{,}536 + 1 = 56{,}537$, which he knew was prime as well. That gave him five Fermat primes using his version of the formula where $x = 0, 1, 2, 3$, and 4, resulting in the prime numbers: 3, 5, 17, 257, and 65,537. Fermat himself had gotten this far.

Euler then reported that for $x = 5$, the result of $2^{2^5} + 1$, simplified to $2^{32} + 1$, gave him the result 4,294,967,297. This time Euler was suspicious. Although he knew prime numbers well, he hadn't memorized them this far, but it didn't look prime to him. To prove mathematically that all numbers of the form $2^{2^x} + 1$ are prime would have been very difficult. However, to prove that they are not all prime required only one counterexample, which Euler was able to find. As he played with it, he discovered that the number 641 is a factor—$641 \times 6{,}700{,}417 = 4{,}294{,}967{,}297$. Euler had his counterexample. Euler now had a proof that Fermat's statement was wrong.

In fact, over the years, no one has ever found a Fermat prime bigger than 65,537. Apparently Fermat was simply lucky that his first five Fermat numbers were prime.

Another well-known problem that had baffled the Bernoulli family was known as the Basel Problem. It asked what is the sum of the reciprocals of the perfect squares: $1/1 + 1/4 + 1/9 + 1/16 + 1/25 + \ldots$? In 1689, then 35-year-old Jacob Bernoulli brought the problem to the attention of the mathematicians of Europe: "If anyone finds and communicates to us that which thus far has eluded our efforts, great will be our gratitude." Jacob knew that it converged to approximately 8/5, but so far neither he nor any other mathematician had been able to find its exact sum. Euler found the sum 46 years later in 1735 when he was 28 years old: the surprising solution is $\pi^2/6$, although it may seem odd that the number π is involved, since the expression clearly has nothing to do with circles. Johann Bernoulli, Euler's mentor, was impressed: "If only my brother were alive!" he said fervently. "We knew it had to be a significant number, but neither Jacob nor I could find it! Hallelujah!"

Euler's solution to that famous problem convinced everyone in mathematics that he was a formidable mathematician, although they would not be truly satisfied until Euler justified his assertion with a formal proof. They didn't question whether Euler had the exact sum—they knew he was right—but they wanted a proof, and Euler was able to supply later several different proofs.

Another topic that Euler explored was perfect numbers, which had fascinated the ancient Greeks. Euclid had defined a perfect number as a number that is the sum of its proper divisors—all its divisors except itself. For example, the number six is the first perfect number since its proper factors are 1, 2, and 3 and $1 + 2 + 3 = 6$. The number 28 is the second perfect number since its factors are 1, 2, 4, 7 and 14: $1 + 2 + 4 + 7 + 14 = 28$. Euler had first learned about perfect numbers as a boy when he had read the *Coss* with his father.

Marin Mersenne, one of Fermat's correspondents, found that a number of the form $2^n - 1$ (in other words, a number that is one less than a power of two) might be a prime number if the number n itself is prime, although he was aware that not all such numbers are prime. If such a number is prime, it is a Mersenne prime, and each time that a new "biggest" prime number is discovered today, it is a Mersenne prime. The difficulty is to prove that the resulting enormous number really has no factors other than itself and one, since the numbers that are currently being discovered are impossibly big.

In 300 B.C., Euclid stated that every even perfect number had to be of the form $2^{n-1}(2^n - 1)$, where $2^n - 1$ is a Mersenne prime, although Euclid could not have written it in this form since he didn't have access to literal algebra. He would have had to write it in words instead. Although Euclid was unable to prove it, Euler was able to present a satisfactory proof in 1750. William Dunham has suggested calling it the Euclid-Euler Theorem.

Euler also worked on amicable numbers, pairs of numbers where each of the numbers is the sum of the other number's proper factors. The numbers 220 and 284, the smallest pair of amicable numbers, were known to Pythagoras and his followers in about 500 B.C., since the factors of 220 (1, 2, 4, 5, 10, 11, 20, 22, 44, 55, and 110) add up to 284, while the factors of 284 (1, 2, 4, 71, and 142) add up to 220.

When Fermat found another pair of amicable numbers—17,296 and 18,416—in 1636, Descartes, who considered Fermat a competitor, took that as a challenge and found still another amicable pair—9,363,584 and 9,437,056. Those three pairs of amicable numbers were the only ones known until Euler took up the challenge. Using a formula that he devised, he found another 60 pairs!

Over the years, Euler spent a considerable amount of time working on number theory, which has since come to be regarded as an important part of mathematics, in great part due to Euler's work in the field. Four entire volumes of the 70+ volume collection of his writings are on the subject. His work on number theory, which was aided and abetted by Goldbach, brought number theory to the attention of serious mathematicians. Carl Friedrich Gauss, the greatest mathematician of the next generation after Euler, made the famous statement, "Mathematics is the queen of science, and ... [number theory] ... is the queen of mathematics." Euler would have agreed.

37
Magic Squares

In his spare time, Euler occasionally played with the popular topic of magic squares, which had been explored first by the Chinese beginning about 2200 B.C. A 3 × 3 magic square is the simplest nontrivial magic square, using each of the digits 1–9 exactly once in such a way that the sum of each row, each column and each diagonal is the same number (see Figure 8). In the case of this three × three square, the sum is 15.

Albrecht Dürer included a 4 × 4 magic square in his famous etching *Melancholia* in the year 1514, cleverly including the date of its construction in the two center squares of his bottom row. Not only do the rows and columns all add up to 34; the four numbers in

4	9	2
3	5	7
8	1	6

Figure 8. A 3 × 3 magic square.

16	3	2	13
5	10	11	8
9	6	7	12
4	15	14	1

Dürer's magic square from *Melancholia*.

52	61	4	13	20	29	36	45
14	3	62	51	46	35	30	19
53	60	5	12	21	28	37	44
11	6	59	54	43	38	27	22
55	58	7	10	23	26	39	42
9	8	57	56	41	40	25	24
50	63	2	15	18	31	34	47
16	1	64	49	48	33	32	17

Ben Franklin's magic square.

each of the four corners also add up to the same sum of 34. Careful study of Dürer's square reveals several more sums equal to 34.

The American statesman and scientist, Benjamin Franklin, a contemporary of Euler, produced his own 8 × 8 magic square when he was a boy. Franklin's square has the magic number of 260 as the sum of each row and each column. Additionally, each half-diagonal when it is bent back at an angle of 90° in the opposite direction for a second half-diagonal also has the magic sum of 260. Franklin considered it child's play, but his result is intriguing. A recent biography of Franklin by Joyce E. Chaplin identifies Franklin as *The First Scientific American.*

1	48	31	50	33	16	63	18
30	51	46	3	62	19	14	35
47	2	49	32	15	34	17	64
52	29	4	45	20	61	36	13
5	44	25	56	9	40	21	60
28	53	8	41	24	57	12	37
43	6	55	26	39	10	59	22
54	27	42	7	58	23	38	11

Leonhard Euler's magic square involving the knight's move.

Although Fermat and Descartes played occasionally with magic squares, their work with them was not noteworthy. Leonhard Euler, however, brought the puzzle to a new height. His most famous 8 × 8 magic square not only has all rows and columns that add up to the magic sum of 260; each quarter of the square is also a magic square in itself with each row and column adding up to 130. As if that were not enough, as Euler constructed his magic square, he began with the number one in the upper left corner, and then he generated the square by placing the numbers consecutively in the knight's move in chess, moving up or down two and to the right or left one, or up or down one, and to the right or left two, throughout the whole square. As a good chess player, Euler included the knight's move as an amusing wrinkle to his square.

A typical Sudoku puzzle is nothing more than a 9 × 9 magic square with nine smaller 3 × 3 magic squares in each ninth of the puzzle, although the diagonals do not add to the magic sum. The Sudoku solver's challenge is to fill in the blanks. The constraints that Dürer, Euler, and Benjamin Franklin put on their squares were far more demanding than a standard Sudoku.

38
Catherine the Great Invites Euler to Return to St. Petersburg

Although Euler had been charmed at first with his life in Berlin, far from the intrigues of the Russian court, he soon realized that working with Frederick the Great was problematic. If Frederick the Great had been able to accept Euler as a scholar whose daily life was irrelevant to their relations, Euler might have continued happily in Berlin for many years. However, during his 25 years in Prussia's capital, Euler was repeatedly annoyed by Frederick's attitude toward him. His monarch viewed Euler—without a doubt Europe's most distinguished mathematician—as little more than a lowborn, common peasant who happened to be good with numbers. Frederick knew that Euler had accomplished many impressive things for him, but the fact remained that Euler was totally unsophisticated in his tastes. For example, Frederick could not understand why the man still preferred to speak German although he could easily have switched over to the elegant French language! Frederick was dismayed by Euler's openly devout lifestyle, in which he displayed a humility that was incompatible with the fashions of Frederick's royal court. By now, these two remarkably capable men, who were in fact dependent on each other in many ways, had lost any mutual ground. To Euler, St. Petersburg looked better and better.

During Euler's many years at the academy in Berlin, Frederick belligerently refused to recognize him as the head of his academy.

When Frederick tried to lure d'Alembert to Berlin, d'Alembert was surprised. Was Frederick actually suggesting that Euler was a lesser scholar than d'Alembert? Regardless of Frederick's opinion, everyone else in Europe—including d'Alembert—knew Euler's worth. Although Frederick implied that he preferred d'Alembert's mathematics to Euler's because Euler's was beyond his understanding, Frederick actually couldn't understand d'Alembert's mathematics either. In fact, Frederick claimed to resent Euler's abstract mathematics only because his mathematics seemed like an easy target. D'Alembert's only real advantage was that he was a charming Frenchman—who admittedly had two functioning eyes—and Euler wasn't.

Beginning in 1761 when he was 54 years old, Euler began to consider seriously returning to St. Petersburg. At one time, Frederick the Great had asked Euler how he had become so knowledgeable in science and mathematics. Euler had replied that it had been his good fortune to have spent many years at the Academy of St. Petersburg, which gave him the opportunity to study and create in that supportive, scientific community. Frederick the Great was baffled by Euler's answer. Frederick would probably have abhorred such a life in St. Petersburg.

When Euler sold the estate he had bought for his mother in Charlottenburg, he was presumably clearing the way for his eventual departure from Berlin. Euler confided to a friend that if he could not arrange to return to St. Petersburg, he might simply retire to the Helvetian Confederation, falling back on his accumulated wealth which had resulted from his work for both St. Petersburg and Berlin as well as from his prize money won through numerous scientific contests. Ironically, he had also recently won a surprisingly large sum from the lottery, which he had entered purely out of curiosity. He commented facetiously that that winning was almost as good as winning the Paris Prize again! Euler also considered moving to England, although nothing came of either of those ideas.

In July 1763, the new Tsarina, Catherine the Great, who was eager to return the St. Petersburg Academy to its earlier status, began to work seriously toward bringing Euler back to St. Petersburg, although the move would not actually happen for another three years. War between Russia and Prussia had just ended in 1763, but relations between the two kingdoms were still strained.

By 1765, Euler had begun serious correspondence with the Russians for his return to Russia. Having been humiliated by Frederick the Great for so long, Euler had no intention of putting himself in a similar position in the Russian court. Catherine the Great, who was willing to be generous, allowed Euler to strike a hard bargain. Euler, well aware that the presidency of the academy was a formal position reserved for a nobleman of the court, demanded the position of vice president of the academy and a salary of 3,000 Rubles for himself, a widow's pension of 1,000 Rubles for his wife, a full professor status for his son Albrecht at an annual salary of 1,000 Rubles, a guaranteed position in the academy for his son Karl (who was then a budding mathematician), and a position as an officer in the Russian military for his son Christoph. Additionally, he wanted to have a staff of young mathematicians to work with him, as well as free housing, heating, and guaranteed release from ever quartering Russian soldiers in his house.

Catherine the Great's response was enthusiastic: The great mathematician should have almost everything he wanted—everything except the vice presidency. However, that was not intended as a slight. She had recently given another scientist the title of vice president, and she argued that Euler shouldn't be put at the same level as anyone else—Euler's status was infinitely higher. She recognized him as the noblest of scientists, giving him 10,800 Rubles to buy a house including furniture, and she gave him the full-time services of one of her cooks. His assignment was simply to return her academy to the splendor that Peter the Great had envisioned for it.

The next step was for Euler to request Frederick to release him from his contract in Berlin. Euler wrote to Frederick in February 1766 but received no reply. Frederick ignored Euler's second letter also, but his third letter was answered with a direct statement: Stop bothering me. Frederick had no intention of losing the world's greatest mathematician from his court, regardless of how he might belittle Euler in his private dealings with him. Euler politely asked once again for permission to leave. This time, Frederick replied tersely that Euler, who was then 59, might go. Catherine the Great, who had staged a successful, bloodless coup over her husband just a few years earlier, was a distant cousin of Frederick, and Frederick decided that it was unwise to further antagonize that powerful woman.

Euler easily arranged to travel to Russia with his two older sons and his daughters plus various other relatives and servants, but negotiating with Frederick for the release of his youngest son Christoph from his position in the Prussian army was more difficult. In fact, the rest of the family had to leave without Christoph, vowing to do what they could for him once they arrived in St. Petersburg. When the formidable ruler Catherine the Great later gave the order, Christoph was allowed to join his family in Russia as well.

The Eulers, who traveled to St. Petersburg by carriage, were given a near royal welcome wherever they went. For ten days they were lavishly entertained in the court of the king of Poland. Then they spent four days in the court of the Duke of Courland, outside Riga, Latvia, where they were regally entertained once again. The Eulers had sent many of their possessions back to St. Petersburg by ship, but unfortunately that ship sank in the Baltic, causing them to lose many of their personal possessions. When Frederick the Great heard of the shipwreck, he commented cynically that it must have cost Euler many mathematical formulas. When the entourage finally arrived in St. Petersburg, they were received as conquering heroes.

By this time, Euler was grateful to arrive in St. Petersburg, and to remain there for the rest of his life. The cataract that had been developing in his left eye—his "good" eye—had become so clouded that he was able to see almost nothing. He was functionally blind. In desperation, Euler decided to have the cloudy lens removed surgically from the eye—a risky procedure—but by that time it seemed that he had little to lose. At first, the surgery was successful: Euler could see! Unfortunately, however, complications arose just a few days later, and the resulting infection caused Euler dreadful pain and ultimately left him almost completely blind. He could see light and dark, but little more. He could neither read nor write nor recognize the faces of his family and friends.

Such a blow might have ended any other scientist's career, but once again Euler accepted it and moved ahead with his research and writing. He requested a table to be built in his work room with a large slate on its top surface. Euler worked with his helpers, to whom he dictated all that he wrote, and when he needed to initiate a new

Euler as an older man.

symbol (he was still the pioneering mathematician in the world), his helper would copy the symbol from the slate into the text. Euler often walked round and round the table as he worked, dragging his left hand along the surface. In later years a shiny path was visible where his fingers had swept.

Shortly before this time, Euler dictated his famous algebra textbook in the German language: *Vollständige Anleitung zur Algebra* [*Elements of Algebra*]. His assistant on this project was a young man whom he had brought back with him from Berlin to serve as a domestic servant to the family. The young man knew no mathematics beyond simple arithmetic when they began, but through their cooperative work, at the end of the volume the assistant was said to be completely competent in all of algebra.

Much of Euler's work, even in his old age, involved lengthy calculations, which he preferred handing off to his helpers, claiming that he couldn't keep track of such detailed work in his head. However, on at least one occasion, Euler disagreed with the answer his helpers gave. Euler then refigured the amount entirely in his head. When his helpers checked their work again, they had made a mistake. Euler was right. It was clear that Euler wasn't really delegating the calculating—his helpers were providing nothing more than a backup for him.

A disaster occurred in 1771 when Euler was 64 years old, justifying Katharina's worst fears: the Eulers' house (along with 550 others in the city) burned to the ground. Euler was hard at work in his room at the time, but his handyman Peter Grimm, another expatriate from Basel, rushed into the burning house, risking his own life, and carrried the great man over his shoulder to safety. A Russian nobleman, Count Vladimir Orlov, rescued Euler's current manuscripts at the same time. When Catherine the Great learned that Euler's house had burned, she immediately gave him 6,000 Rubles to rebuild the house, which was promptly accomplished, only this time Euler's wife Katharina was relieved to have a handsome structure of stone instead of wood.

In 1773, Euler wrote to his friend Daniel Bernoulli in Basel, asking him to recommend an assistant for him from Basel. Despite his competence in many languages, Euler still preferred to work in German—preferably German with a Swiss accent—which had become more important now that all his work needed to be oral. Daniel sent a young scholar named Nicholas Fuss from Basel to live with Euler and to serve as his personal secretary. Fuss worked closely with Euler for ten years, eventually marrying Euler's granddaughter, the daughter of his oldest son Albrecht.

In 1773, only two years after the fire, Euler was devastated when his wife Katharina died at the age of 67. Katharina had cared for him, and he for her, for many years. Three years later he remarried, this time the 53-year-old half-sister of his first wife Katharina, in a private ceremony in the family's living room. Although his sons had expected that he would simply move in with one of them, Euler preferred to be independent. He and Salome Abigail Euler lived happily together for seven more years.

Euler was still a formidable scientist at this time, producing more than 400 separate works in mathematics and physics in the 17 years after his return to St. Petersburg. In spite of his blindness, he was still the most productive mathematician in Europe even then.

In 1783, Catherine the Great made the Princess Catherine Romanova Dashkova, sometimes called "Catherine the Little," the director of the academy. Although she was no scholar, she had great respect for scholars and their work, and she humbly accepted the position and carried out her duties well. When she made her first entrance into the Academy, she chose to have as her escort Euler, blind yet still the greatest mathematician in the world. He had to be led into the hall by his assistant Fuss, who was appalled to find that someone else had already taken the seat of honor, which should have been reserved for Euler, next to the princess. When she saw what had happened, she maintained her composure and announced confidently to the academy: "No matter! Professor Euler, you may sit wherever you please. That seat will automatically be the most important seat simply because of your presence." Fuss then led Euler to another seat, and the academy began its meeting.

The princess later arranged for a mural to be painted on the wall of the assembly room of the academy, celebrating the wisdom of geometry, including many of Euler's discoveries, formulas, and drawings. It was intended as a fitting tribute to the great man.

39
The Basel Clan

In 1780, 10-year-old Leonhard Euler, the son of Karl Euler, said to his 73-year-old grandfather Leonhard Euler, "Grandfather, you didn't live in St. Petersburg when you were a boy, did you?"

"No, I grew up far from St. Petersburg," the renowned mathematician said. "I am a native of Basel in the Helvetian Confederation. I grew up in the village of Riehen, an hour's walk outside of Basel."

"What was it like?" young Leonhard asked. "I can't imagine living anywhere else than St. Petersburg."

"I had the most marvelous childhood imaginable," Euler said. "My father was a pastor in the Reformed church in our village, my mother was a wonderful and wise woman, and I had two charming younger sisters and a younger brother. We had a wonderful life together. I have to admit, though, that my biggest advantage was that my father was a friend of Professor Johann Bernoulli at the university in Basel. There is no doubt that it was Bernoulli who gave me the background that I needed to become the mathematician I am today."

"But Bernoulli wasn't more important than you are, Grandfather, was he?" young Leonhard asked.

"Johann Bernoulli was the most important mathematician of his time," Euler said. "People called him the Prince of Mathematicians."

"But isn't that what they call you?" young Leonhard asked.

"There's no comparison!" Euler said. "I'm sure Bernoulli was many times more important than I am. But before I tell you about him, I should tell you first about his older brother Jacob. My father—your great-grandfather—studied under Jacob Bernoulli, and he never forgot that brilliant man. My father was determined that I should experience the wonders of mathematics in the same way that he did as a student."

"And did you?" the boy asked.

"Oh, yes!" Euler said. "When I was a boy, Jacob Bernoulli's younger brother, Professor Johann Bernoulli, generously offered to be available to me on Saturday afternoons so that I might ask him questions about mathematics as I encountered them in my studies. He was as kind to me as a man could be, and for him to devote all that time to me—an ignorant young student—was probably due to my father's influence."

"So did you have to walk all the way to Basel for those meetings?" young Leonhard asked.

"During the week at that time," Euler explained, "I was living with my grandmother in town. That is where the university is. I walked home to our village each Saturday after my meeting with Professor Bernoulli—and that was a long, boring walk, I can tell you—and then I walked back to Basel on Sunday afternoon for the next week of classes."

"Did you like staying with your grandmother?" Leonhard asked.

"Oh, yes," Euler said. "She was a smart woman, and we had great times together. You would have liked her."

"Why did you leave Basel?" the boy asked. "Couldn't you have been a professor there?"

"Well, Professor Bernoulli held the only chair in mathematics there at the time," Euler explained. "If I wanted to be a mathematician, I had to go somewhere else to do it. St. Petersburg was able to offer me many advantages, including the friendship of Daniel Bernoulli when I was first here. Daniel was the second son of my professor, and we had splendid times here working together."

"I've heard of Daniel Bernoulli," Leonhard said. "He is a very important scientist, isn't he?"

"Yes, and he is a very dear friend," Euler said. "Daniel and I worked together here in St. Petersburg for several years before he returned to Basel, and his support in those first few years was critical to me."

"Is he back in Basel now?" the grandson asked.

"Yes, he never liked St. Petersburg," Euler said.

"Really?" Leonhard asked. "But St. Petersburg is such a beautiful place to live!"

"Yes, it is," Euler said. "I can't explain why, but he hated it here in St. Petersburg. He always dreamed of returning to Basel, and when he finally was called to a position at the university there, he didn't hesitate."

"Then why didn't you return to Basel, too?" the child asked.

"Well, I have always loved working at the academy here in St. Petersburg," Euler said. "Through all these years, there have been many brilliant scientists here, all working together, supporting and challenging each other. I found that I work best here."

"Will I ever get to meet your friend Daniel Bernoulli?" young Leonhard asked.

"Probably not," Euler admitted. "By now, he is an old man like me, and I doubt that he has any desire to travel to St. Petersburg again."

"And you don't plan to go back to Basel again either, do you, Grandfather?" his grandson asked.

"No, I considered it several times," Euler said. "I was asked to return to Basel when my mentor Johann Bernoulli died, but it didn't seem wise at the time. The political situation in Russia was calm, and my work here was going well. I have sometimes questioned that decision—I certainly regretted it toward the end of my time in Berlin as I struggled to deal with Frederick the Great."

"You knew Frederick the Great?" Leonhard asked.

"Yes, but that was not a happy time in my life," Euler said. "Frederick the Great didn't want a pure mathematician—he wanted someone who could be charming company. I'm afraid that has never been my strength. I think it would be safe to say that Frederick the Great didn't like me very much. And your uncle Christoph, my third son, suffered at his hands as well. He was in the Prussian army when I finally got permission for the rest of the family to leave Berlin, and Frederick would not allow Christoph to leave the Prussian army. He had to wait until the rest of us had arrived in St. Petersburg, and then Catherine the Great used her influence to arrange his release from Frederick's army."

"She is our empress, isn't she?" young Leonhard asked.

"Oh, yes!" Euler said fervently. "She is a grand lady and a fine monarch."

"But you didn't like Frederick the Great so much, did you?" Leonhard asked.

"Frederick the Great could be a difficult man," Euler said.

"But why didn't he like you?" young Leonhard asked.

"He didn't understand my mathematics," Euler said, "and mathematics is the only thing that I do well."

"I'm sure that isn't true," his grandson said. "Everyone here says you are the most important scholar in the world. I've heard them say it many times. And you are one of my very favorite people."

"Thank you," Euler said. "I've made some important discoveries, but I am one of many who have done that."

"But you wrote the arithmetic textbook that I am learning from now in school," Leonhard said.

"Are they still using that?" Euler asked. "I wrote that many years ago."

"My copy of the book is old, but it is still good, and I love it," Leonhard said. "Did you know that mathematics is my favorite subject in school?"

"I'm pleased to hear that, young man," Euler said. "There is no subject that is more important. You must never stop learning it. No matter how much of it you have learned, there is always more to master. It is beautiful."

"Will I always find textbooks that you wrote to guide me through?" Leonhard asked.

"When you run out of my books," Euler said, "then it will be time for you to write books for the next generation of students. I truly hope that you will do that."

"When I grow up, will I be able to travel to Basel?" Leonhard asked. "There is a river there, isn't there?"

"Yes, the Rhine River flows through the center of town," Euler said, "but it is totally different from the placid Neva. The Rhine practically roars through Basel."

"I'd like to see it," Leonhard said. "It would be accurate to say that my roots are there, wouldn't it?"

"Yes, we are Swiss even though we haven't lived there for many years," Euler said. "I am who I am because of my upbringing in Basel, and that is part of your background, too. However, I must credit St. Petersburg, too—it was in this utopia for scholars that I matured into the scientist that I am today. You are very fortunate to be growing up here."

"I think so too, Grandfather," young Leonhard said.

"Basel has had an unusually strong impact on the development of mathematics," Euler explained, "with Jacob and Johann Bernoulli, then Johann's sons Nicolaus, Daniel and Johann Bernoulli, and quite possibly some of their children as well. That's quite a record for one small city in the Helvetian Confederation."

"Don't forget to mention yourself, Leonhard Euler, the Prince of Mathematicians!" young Leonhard said.

Index

A

Acta Eruditorum, 67, 74, 82, 83, 89, 92, 97, 98, 99, 104, 106, 111, 152, 161
Aeneid, 107, 134
aether, 67, 146
Alba, Duke of (1507–1582), 3, 5, 6
algebra, 10, 11, 45, 55, 59, 62, 78, 100, 119, 139, 141, 145, 223, 242, 249, 260
almanac, 230–231, 233, 237
amicable numbers, 250
Analyse des infiniment petits, 91
analytic geometry. *See* algebra
Anno Mirabilis, 77
Antwerp, 1–5, 7, 8, 28
Arabic, 42
Archimedes (287–212 B.C.), 78, 205
Aristotle (384–322 B.C.), 112
Ars Conjectandi, 110–115, 149
astronomy, 31, 59–60, 63, 193, 223, 239

B

Bach, Johann Sebastian (1685–1750), 188–189
Balaton, Lake, in Hungary, 35
barometer, 67, 70
Barrow, Isaac (1630–1677), 64, 89
Bartholomew's Day, St., 5
Basel Problem, 248–249
Berlin, 198, 219, 220, 224, 225–237, 255–257, 260, 265–266
Berlin Academy, 117, 159, 198
Bernoulli Family Tree, 2, 118
Bernoulli, Daniel (1700–1782), 63, 110, 117, 118, 119, 121, 123–127, 129–131, 145, 149–155, 158, 159, 163–168, 169–176, 177–180, 181–188, 191–199, 201–205, 207, 209, 210, 213, 215–216, 224, 229, 240, 261, 264–265, 267
Bernoulli, Daniel (1751–1834), 118, 198
Bernoulli, Francina (?–1615), 1–5, 7
Bernoulli, Jacob (?–1583), 1–5, 7, 28

Bernoulli, Jacob (1598–1634), 2, 7, 8
Bernoulli, Jacob (1654–1705), 8–11, 13–19, 21–26, 27–32, 33–38, 39–45, 47–52, 53–60, 61–64, 65–76, 82, 83–84, 87–94, 97–101, 104–108, 109–115, 117–119, 121–122, 133, 139, 146, 149, 152, 153, 154, 161, 162, 177, 178, 191, 201, 204, 205, 242, 248–249, 264, 267
Bernoulli, Jacob (1759–1809), 198
Bernoulli, Johann (1667–1748), 8, 21–26, 39, 64, 72–75, 84, 87–95, 98–101, 103–108, 110, 115, 117–126, 133, 143–148, 149–154, 158, 161–164, 167, 171–176, 192, 193, 197–199, 201, 205, 211–213, 224, 229, 240, 242, 243, 249, 263–267
Bernoulli, Johann (1710–1790), 118, 173, 192–196, 198–199, 201, 224, 229, 267
Bernoulli, Johann (1744–1807), 118, 198
Bernoulli Law, 174
Bernoulli, Leon (?–1561), 2
Bernoulli, Nicolaus (?–1608), 2, 7
Bernoulli, Nicolaus (1623–1708), 8–10, 21, 22, 27–30, 45, 54, 65–67, 87, 118, 146
Bernoulli, Nicolaus (1662–1716), 22, 24, 71, 110, 115, 118, 177
Bernoulli, Nicolaus (1687–1759), 115, 177, 179
Bernoulli, Nicolaus (1687–1769), 71, 110, 115
Bernoulli, Nicolaus (1695–1726), 94, 103, 118, 119–122, 123–125, 145, 151–155, 158, 166, 172–173, 177, 181–183, 191, 197, 202, 267
Bernoulli, Verena (1685–1768), 71
blacksmith, 49, 50
blind, 30–31, 35–38, 39–45, 79, 111, 216–217, 220, 258–261
Bouguer, Pierre (1698–1758), 161
Boyle, Robert (1627–1691), 63, 70
brachistochrone, 104–106
Brucker, Margaretha (1678–1761), 133, 141–142, 148, 234, 263
Brunswick, Royal House of, 115

C

calculating machine, 81
calculation of variations, 106
calculus, 61, 64, 77–85, 88–95, 97–100, 104–108, 109, 123, 134, 145, 159, 172, 184, 193, 201, 213, 223, 238, 240, 242–243
Campani, Giuseppe (1635–1715), 63
capillary action, 69–70
Cardano, Girolamo (1501–1576), 30, 45, 111
carpenter, 35–36, 130
Cartesian. *See* Descartes, René (1596–1650)
Carthage, 107
Cartier, M. *See* carpenter

catenary, 97–101
Catherine the Great (1729–1796), 184, 256–258, 260, 261, 266
Catherine the Little (Princess Catherine Romanova Dashkova) (1743–1810), 261
Catholic. *See* Catholic Church
Catholic Church, 3, 5, 59, 127
censors, 127, 165
chance. *See* probability
Châtelet, Emilie Marquise du (1706–1749), 198–199, 226
chronometer, 130, 131
Church of Rome. *See* Catholic Church
cipher, 53–54, 82
cloth, life of. *See* pastor
Cocles, Horatio, 95
code. *See* cipher
Collection. *See* Pappus (ca. 290–350)
combinations, 112
combustion, 63, 70
comet, 59, 60, 67
Condorcet, Nicolas de (1743–1794), 125
Coss, 10–11, 13–18, 44, 55–56, 139–141, 249
cossists. *See Coss*
curve of quickest descent. *See* brachistochrone
cycloid, 97–98, 106

D

d'Alembert, Jean (1717–1783), 229–230, 256
derivative, 82
Descartes, René (1596–1650), 51–62, 71, 73–74, 78, 83, 89, 90, 145–146, 246, 250, 254
De Witt, Johann, 114
diamond, 5
Dido, Princess, 107
digit, 37, 39–43, 139
Diophantus (250 A.D.), 246
du Châtelet, Emilie Marquise (1706–1749). *See* Châtelet, Emilie Marquise du (1706–1749)
Dunham, William (1947–), 249
Dürer, Albrecht (1471–1528), 251–254

E

Elements of Algebra, 260
Elizabeth. *See* Waldkrich, Elizabeth
Euclid (325–265 B.C.), 78, 249
Euler, Johann Albrecht (1734–1800), 209, 213, 234, 257, 261
Euler, Leonhard (1707–1783), 72, 78, 84, 133–137, 139–142, 143–148, 154, 155, 157, 159, 161–168, 169–176, 177–180, 181–189, 191–195, 201, 203, 205, 207–213, 215–217, 219–224, 225–236, 237–243, 245–250, 251–254, 255–262, 263–267
Euler, Paul (1670–1745), 72, 84, 133, 137, 139–142, 144, 146–148, 224, 234, 249, 263
even functions, 183
Exercitationes mathematicae, 127, 149
exponent, 55, 59

F

Falkner, Dorothea (1673–1764), 93–94, 103, 108, 117–122, 124–125, 197
family tree, Bernoulli, 2, 118
faro, 149–150
Fatio-de-Duillier, Jean Christophe (1664–1753), 71, 90
Fatio-de-Duillier, Nicolas (1664–1753), 90
Fermat, Pierre de (1601–1665), 106, 111, 245–250, 254
Finow Canal, 237
fire, 209, 220, 226, 260
Flamsteed, John (1646–1719), 63
fluid dynamics, 150, 169–172, 174–176, 194, 202
fluxions, 78–85, 95, 105, 213, 243
Frankfurt, 4–6, 7, 28, 164, 234
Franklin, Benjamin (1706–1790), 231, 253–254
Frederick the Great, King of Prussia (1712–1786), 198, 219, 223–224, 225–235, 237–242, 255–258, 265–266
Frey, Maria (1596–1625), 8
function, 82, 220, 242
Fuss, Nicholas (1755–1825), 261–262

G

Galileo Galilei (1564–1642), 98, 99, 127, 238
Gauss, Carl Friedrich (1777–1885), 78, 250
genius, 59, 80, 81, 101, 148, 162, 171, 211, 226
geography, 51, 165, 209, 217
geometric series, 14
Geometrie. *See* Descartes, René (1596–1650)
geometry, 18, 25, 59, 61, 64, 67, 71, 78, 83, 89, 90, 100, 144, 145, 150, 187, 223, 239, 242, 262
gnomon, 47, 49, 50
Goldbach, Christian (1690–1764), 123, 126, 151–154, 163, 164, 209–211, 217, 245–250
gold pieces, 5
gout, 48–49, 58, 119
graph theory, 187, 240–241
gravity, 67, 79, 97, 100, 104, 170
Greenwich Royal Observatory, 63, 131
Groningen, 94, 103, 104, 108, 117, 120, 121, 153
Gsell, Katharina (?–1773), 207–209, 215, 220–222, 226, 234, 260, 261
gunnery, 237–239

H

handicap, 113
Harrison, John (1693–1776), 130–131
Helios, 31
Helvetian Confederation, 8, 48, 162, 165, 203, 256, 263, 267
Hermann, Jacob (1678–1733), 153, 168, 183, 195, 209
Hippocratic oath, 3
Hooke, Robert (1635–1703), 63–64, 70

Hôpital, Marquis de l' (1661–1704), 90–91, 100, 104–106
Horatio. *See* Cocles, Horatio
hourglass, 130, 150
Hudde, Jan (1628–1704), 61–63
Huguenot, 1–6, 7
Huygens, Christian (1629–1695), 71, 78, 81, 83, 90, 93–94, 97, 98, 111, 112, 129, 161
Hydraulica, 176
Hydrodynamica. *See* fluid dynamics
hydrodynamics. *See* fluid dynamics
hyperbolic geometry, 100

I

infinitesimal, 78, 84
insurance, 111–113
integral calculus, 97, 243
integrating factor, 184
integration, 243
Introductio, 241, 242
isochrone, 97–98, 161
isoperimetric problem, 107, 241

J

Jan Suratt (fictional), 2, 4
Journal des Sçavants, 93, 110
Justus de Boer (fictional), 1–4

K

Königsberg Bridges, 184–187
Korff, Johann Albrecht (1697–1766), 209
Krafft, Georg Wolfgang (1701–1754), 209

L

Large Numbers, Law of, 113–114
Latin, 18, 21, 32, 38, 51–52, 56, 59, 62, 63, 80, 82, 109, 110, 120, 121, 134, 139, 143–145, 161, 166, 181, 184, 194–195, 210, 211–212, 212, 238, 242, 245, 246
Latin school, 141, 143
Leibniz, Gottfried (1646–1716), 77, 80–85, 88–90, 92, 95, 97–100, 104–106, 113–115, 117, 157–159, 172, 193, 212–213
Lettres à une Princesse d'Allemagne, 239–240
licentiate, 21, 27, 88, 121
literal algebra. *See* algebra
Livy, 95
longitude, 129–131, 202
Lostanges, Marquis de, 47
Luther, Martin (1483–1546), 65–66

M

magnetism, 201
Malebranche, Nicolas (1638–1715), 53–59, 67, 90
marquis. *See* Lostanges, Marquis de
mason, 49, 50
master's degree, 9, 27, 145, 161
math gene, 8
Mattmüller, Martin, 109
Maupertuis, Pierre-Louis Moreau de (1698–1759), 193, 198–199, 229, 235
maxima, 64, 82

Mechanica. See mechanics
mechanics, 68–71, 152–154, 162, 164, 168, 187, 213, 223, 239
Meditationes, 51
mercury, 69, 130
Mersenne, Marin (1588–1648), 246, 249
Mersenne prime, 249
Michelotti, Pietro Antonio (1673–1740), 126–127, 172
Micrographia. See Hooke, Robert
minima, 64
ministry. *See* pastor
Mint, Master of the, 80
Monmort, Pierre Rémond de (1678–1719), 98–101
moral certainty, 112–114, 180
moral value, 180
Morgagni, Giovanni Battista (1682–1771), 126
mortality statistics, 113–115
motto, 31–32, 59
Music, 187–188, 235, 239

N

nature or nurture. *See* math gene
navigation, 129
negative, 62
Newton, Isaac (1643–1727), 75, 77–85, 89–90, 95, 97, 104–106, 117, 146, 159, 170–172, 198–199, 202, 212–213, 243
notation, 55–56, 82, 84, 106, 188–189, 242–243
number theory, 223, 245–250

O

odd functions, 183
Oldenburg, Henry (1618–1677), 212
omen. *See* comet
optics, 78–79, 106, 239
orbit, 59–60, 196

P

Pappus (ca. 290–350), 17–19, 21–26, 44, 57
parabola, 98, 99
Paris Academy, 100–101, 117, 157, 159, 204
Paris Prize, 127, 130, 150, 161–162, 196–198, 201–202, 213, 256
Pascal, Blaise (1623–1662), 111–113, 246
pastor, 27–30, 65, 88, 120, 134, 146, 148, 263
pendulum, 97, 111, 129, 130, 161, 202
perfect number, 249
Peter the Great (1672–1725), 157–159, 165, 167, 219, 257
Petrovna, Elisavetta (1709–1762), 219, 235
Phaeton, 31–32
physics, 67–72, 100, 162, 166, 172, 187–189, 193, 204, 210, 223, 237–239, 242, 261
place value number system, 37, 41–43
Plato (?–348 B.C.), 10–11, 16, 19, 30
poetry, 51, 109

polar coordinates, 73–75
Pope. *See* Catholic Church
powers of two, 15–16, 245–248, 249
prime numbers, 134
Principia, Newton's, 83, 199
priority dispute. *See* calculus
probability, 45, 110–113, 149
Protestant. *See* Huguenot
Pythagoras (ca. 500 B.C.), 15, 250
Pythagorean Theorem, 246

Q

quill, 57, 69

R

reciprocal trajectory, 181–182
regency, 159, 164, 219
Republic, The. *See* Plato (?–348 B.C.)
respiration, 63, 126
Robins, Benjamin (1707–1751), 237–239
Rudolph, Christoff. *See Coss*

S

Sans Souci, 237
Schumacher, Johann Daniel (1690–1761), 167–169, 208, 223
series, 13–15, 109, 223, 243, 245
Shakespeare, William (1564–1616), 188–189
smallpox, 167, 204
soldiers, quartering of, 220–223
Spanish Fury, 6, 7
spiral, logarithmic, 32, 40
spira mirabilis. *See* spiral, logarithmic
St. Petersburg Academy Journal, 179, 183–184, 210, 224, 225
Stupanus, Judith (1667–1695), 71, 72, 115
Sudoku, 254
sundial, 47–51

T

Theory of Sound, Dissertation on the, 162
Thirty Years War, 8
tides, 51, 201, 223
Torricelli, Evangelista (1608–1647), 70
Tschirnhaus, Ehrenfried Walther von (1651–1708), 88–89

U

ulcer, 155, 173

V

van Schooten, Frans (1615–1660), 52, 61
variable, 11, 55–56, 59, 62, 100
Varignon, Pierre de (1654–1722), 91, 100, 115
variolation. *See* smallpox
Venice, 126–127, 149, 157, 165, 175
Viète, François (1540–1603), 19, 54, 55
vision, 216–217, 220, 235, 258–260, 278–279,

Voltaire (1694–1778), 198–199, 225–228

W

Waldkirch, Elizabeth, 30, 35–38, 39–44
Wallis, John (1616–1703), 64, 78
Wiles, Andrew (1953–), 246

Z

zero, 40–41, 43, 91, 243

Printed in the United States
by Baker & Taylor Publisher Services